INTEGRATED GEOSPATIAL
TECHNOLOGIES

INTEGRATED GEOSPATIAL TECHNOLOGIES

A Guide to GPS, GIS, and Data Logging

Jeff Thurston

Thomas K. Poiker

J. Patrick Moore

WILEY

JOHN WILEY & SONS, INC.

Published by John Wiley & Sons, Inc., Hoboken, New Jersey
Published simultaneously in Canada

For general information on our other products and services or for technical support, please contact our
Customer Care Department within the United States at (800) 762-2974, outside the United States at (317)
572-3993 or fax (317) 572-4002.

Wiley also publishes its books in a variety of electronic formats. Some content that appears in print may
not be available in electronic books. For more information about Wiley products, visit our web site at
www.wiley.com.

Library of Congress Cataloging-in-Publication Data:
    Thurston, Jeff, 1956–
    Integrated geospatial technologies : a guide to
GPS, GIS, and data logging / Jeff Thurston, J. Patrick Moore, Thomas K. Poiker.
    p. cm.
Includes bibliographical references and index.
ISBN 0-471-24409-0 (cloth)
    1. Geographic information systems. 2. Global Positioning System. I. Moore, J. Patrick,
1961– II. Poiker, Thomas K. III. Title.

G70.212.T54 2003
910′.285—dc21   2003041166

Printed in the United States of America

10 9 8 7 6 5 4 3 2 1

CONTENTS

PREFACE

Several geospatial technologies have been developed over the last few decades. Geographic information systems (GISs) have led that path, influencing most other technologies, spatial data collection, and management as well as representation. The integration of spatial technologies revolves around GIS for many applications, research, and education programs. A global positioning system (GPS) is used routinely in many geospatial projects, and when coupled to wireless technologies, provides a means for location-based services (LBSs) and mobility products and services. Other technologies, such as remote sensing, have increased in resolution and availability. Meanwhile, in situ laser instrumentation, light detection and ranging (LIDAR), three- and four-dimensional visualization, and virtual reality environments have entered the sphere of spatial data users and providers. Web mapping has surfaced more recently and is providing new opportunities for cartographic map distribution and representation while becoming more apparent in e-government initiatives.

The integration of spatial technology is seldom discussed holistically. Many users and organizations of spatial technologies voice a common question: How does it all come together? What is the relationship of GIS to GPS? Where can remotely sensed images be used, and what are some of the considerations, such as advantages and disadvantages, compared to other types of technologies? Which data can be used for visualization purposes, and how can these data be acquired? Can my data be used on the Internet, and if so, how?

The purpose of this book is to provide an integrated look at modern spatial technologies and to begin to address some of these questions. This is not a small challenge, since new technologies and applications are growing daily. At the same time, to write about these technologies, it must be recognized that they do not exist without reference to the theory that underpins their use.

We have purposely avoided long theoretical descriptions as well as focusing only on the technologies themselves. There are other books on the market that provide more detailed theory for any given technology and a more exhaustive study by discipline or application. Instead, this book is intended to provide a balanced approach, using an integrative framework, through discussion of theory and technology as they appear in geospatial applications. Consequently, the chapters are written in an integrated fashion, discussing theory and technology as they appear under broader subject headings. We believe that the benefit of this approach will be to cause the reader to think in an integrated manner when using or considering the use of spatial technologies.

We begin the process in Chapter 1 by discussing geotechnology and its integration, identifying the issues and outlining the integration approach. In Chapters 2 and 3 we discuss geodetics and cartography, providing a means of understanding where things are located and how they are influenced by the shape of the Earth. In Chapter 3 we begin to increase the use of examples of integration and discuss direct links to visualization. In Chapter 4 we focus on GIS, the backbone of integration, and discusses some issues relating to GIS applications while providing clear examples of the science of GIS. Chapter 5 is an overview of global positioning systems (GPSs) that provide many examples and factors to be considered when coupling GIS and GPS, as well as outlining the relationship of GPS location to cartographic elements. In Chapter 6 we focus on the integration of technologies, reasoning that earlier chapters have provided enough information for the reader to begin to engage integration conceptually. In Chapter 7 we discuss instrumentation and sensors, a new area that has received little geospatial attention even though digital sensors are a growth area, due to the fact that they couple to other spatial technologies readily and lend themselves to the development of a broad range of applications. Chapter 8 is designed to provide information relating to satellite-gathered spatial information and applications while discussing aerial photography and the delineation and identification of objects for use in GIS. Finally, in Chapter 9 we provide an overview of current visualization technologies, how they link to GIS and GPS, and some consideration of approaches to the representation of spatial information.

We have taken a slightly different approach than that of more conventional chapter-by-chapter approaches to subject matter oriented around a single topic. Although we began by using the traditional approach, it became clear quickly that we could not relate chapters in such a manner as to provide the feeling of integration we wished to express. Subsequently, readers will find themselves referring back (or ahead) through chapters as new information, concepts, and knowledge generate alternative thoughts and new questions arise. This approach yielded the most interesting and robust discussion of integration during writing. It would have been interesting to discuss how politics and economics are coupled to the concepts provided in this book, but we felt that was not necessary to impart an understanding of integrated geospatial technologies.

JEFF THURSTON
BERLIN, GERMANY

PATRICK MOORE
SEATTLE, WASHINGTON

TOM POIKER
VANCOUVER, CANADA

ACKNOWLEDGMENTS

We thank the following people: Ron Sheckler of Integral GIS, Inc., Seattle, Washington, for providing the graphic and appendix item related to Figure 7.5 and the information relating to risk, threat, and security; Bob Parkinson of Calgary, Canada, for information relating to medical uses and epidemiology and the study of foot and mouth disease, a potential scenario in Canada; Dick Puurveen of Edmonton, Canada, for Figures 7.2 and 7.3, relating to digital climate stations from the Breton plots, long-term cropping rotations in western Canada; Ignacio Guerrero of Intergraph Corporation for providing Figures 3.9 and 6.3; Elston Solberg, Forrest Wright, and Rod Kripps for Figure 3.8; the Beaver Regional Sustainable Community Wheel in the province of Alberta, Canada, for their willingness, candidness, and many meetings during the initiation of the sustainable communities project; Bettina Thurston for translation of documents, administration, and support; and Jim Harper of John Wiley & Sons, Inc., for his guidance and positive criticism, together with the Wiley staff members who have contributed to this book.

1

GEOTECHNOLOGY AND INTEGRATION

1.1 INTRODUCTION

Geographic information systems (GISs) and global positioning systems (GPSs) are valuable spatial technologies that affect the lives of individuals, companies, and governments around the world. The natural sciences, business, entertainment, environmental, and transportation studies are all involved. The driving force and benefit of these technologies are related to their ability to provide navigation, location, and analysis of human–human, human–landscape, and human–nature interactions. GPS equipment and services for 1999, as stated by Neal Lane, then an assistant to the U.S. president for science and technology, were projected to exceed $6 billion, approaching $16 billion per year in the first half of the decade 2001–2010. For the year 2000, North American sales of GIS-related products approached $1 billion. The remote sensing industry in North America exceeds $3 billion annually, and wireless technologies and communications for geoscience use are witnessing annual growth rates of about 11 to 19 percent. Several corporations that manufacture GIS software are indicating growth rates of almost 25 percent per annum. Recent advances in wireless communications are permitting many new mobile applications internationally. Sensors and instrumentation are being coupled into these geotechnologies, providing a means to quantify information. The diversity of sensors and instrumentation being coupled to GIS/GPS is expanding continuously.

Answering questions as to where, when, why, and how is not new. What is new is how geotechnology is being applied to address these questions and the rate at which society searches for answers to these questions. As one question is answered, others arise. That is the nature of discovery and exploration, which geotechnologies undeniably enhance. The geotechnology market has grown to include many diverse technologies that did not exist less than a decade ago. Many technologies and techniques applied in surveying, photogrammetry, and remote sensing have a long history. They

1

have been applied effectively and widely over time. Much of the information gathered through their use was not digital as it currently is. This meant long hours in the field to collect the information. It also meant that a large amount of time was needed to process the information. Hardcopy cartography is now being augmented through the use of new geotechnologies. Maps are being generated in real time, and business decisions are being formulated quickly on the basis of such maps. The capture of spatial data using these technologies has resulted in data capture costs lower than those incurred using nondigital methods. Processing of digital technologies also usually results in quicker processing.

New geotechnologies overlap with technologies and methods developed previously. Viewed on a time line (Figure 1.1) there is little doubt that a convergence of technologies is leading to increased need for understanding geotechnology integration. A prominent feature of a time line is education. As technologies appear in the marketplace, a corresponding increase in spatial courses offered has occurred. Traditionally, geoscience professionals have studied geography, surveying, photogrammetry, and cartographic sciences. That is changing as many professionals from outside these fields are now taking courses in GIS, GPS, LIDAR, and other technologies to supplement their knowledge and abilities so that they can use the new approaches more effectively. The technologies appear to act as catalysts, bringing applications and theory together. One person cannot know everything—both theory and technical knowledge. Geotechnologies are therefore often employed in a group manner, involving numerous people from a broad range of disciplines.

Most people are knowledgeable about their discipline but lack an understanding of geotechnologies. Others have sound knowledge of the technologies but lack the ability to apply them effectively to specific disciplines or problems. This discrepancy is a major challenge for the implementation of geotechnologies. Information technology (IT) groups within organizations are rapidly becoming spatially aware. Recent

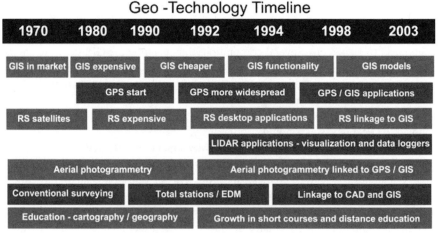

Figure 1.1   Geotechnology time line.

trends incorporate geotechnologies into business applications and organizational computing processes. This is necessitating strategic planning and implementation procedures for these technologies. The costs of geotechnology have dropped considerably. Desktop computing is increasing rapidly in computing performance, and broadband communication capability is expanding. Higher transfer rates, data compression, and ergonomic considerations are leading to improved access to and use of the Internet and spatial data files.

Many geotechnologies have moved from centralized institutions and computing facilities into the consumer market, enabling creative new approaches and providing spatial solutions through both post and real-time applications. This empowers individuals, providing them with alternatives, some of which include higher levels of technology integration. The sharing and communication of spatial information on networks have grown considerably. Geotechnology has shifted from being organization focused to include individuals.

The certification of GIS/GPS and other geotechnology professionals has also become an issue—one not readily solved, due in part to the rapid rise of geotechnologies and the wide variation in individuals and organizations applying them in new and creative ways. The pace has increased so dramatically that licensing and policy bodies can barely embrace all the elements that may be considered within the *geojurisdiction.* Various professionals, including geographers, sociologists, surveyors, environmental and transportation planners, engineers, and business users have differing requirements for spatial data. Accordingly, many of these groups maintain their own set of spatial standards, and groups often overlap on the issue of standards.

These technologies have also been termed *enterprise technologies* in reference to the fact that they will necessarily cross the existing administrative boundaries within organizational structures. The impact on social dynamics within organizations is challenged by these technologies. Internationally, both small and large organizations, as well as governments, continue to embrace geotechnology, striving to develop spatial information initiatives for the new millennium. The European Umbrella Organization for Geographic Information (EUROGI) has recently published a document discussing how geographic information can readily be integrated among agencies in Europe (EUROGI, 2000). The European Territorial Management Information Infrastructure (ETeMII) has produced a white paper discussing transboundary issues related to spatial and other information (ETeMII, 6.2.2, 12/20/01). At the same time, the Canadian government, through its Spatial Data Initiative, has evaluated the impact of spatial technology on the Canadian economy (GeoConnections Policy Advisory Node, 2001). These documents clearly indicate their value in supporting economic growth. Consequently, many spatial information services transcend local applications and are becoming more international in scope. Issues relating to data sharing and metadata have become more important, as have issues regarding accuracy. Geotechnology and spatial information use and applications will continue to grow internationally. This growth will include numerous geotechnologies as individuals and organizations implement new solutions for many spatial issues. That in turn will challenge society to understand these technologies in an integrated manner, requiring new perceptions and knowledge if they are to be applied effectively and shared usefully.

1.2   POSITION DATA

Advances in geotechnology have enabled us to collect, manage, analyze, and present spatial information more quickly. Locating events with higher degrees of accuracy for many applications has previously been unobtainable. Where a forester might accept a location within 100 m, they now seek locations within 1 m. The wildlife specialist might have been able to radio-triangulate an animal's position within a few hectares but now wishes to locate the animal within 10 m or less through the use of GPS animal tracking collars. Oceangoing ships now traverse the planet, charting courses within a few meters through the use of GPS. The ships can now be berthed in harbors using GPS coupled to GIS. Draft beneath a ship is determined through the coupling of electronic depth sensors that are, in turn, coupled to ship position. Costs of producing a 5-m vertical contour map may previously have been hindered due to cost, but have recently become economical with the advent of new geotechnologies. The ecologist can characterize a landscape more quantifiably, the farmer understand soil fertility more easily, and the meteorologist monitor the weather more effectively.

Who would have thought 10 years ago that people would be locating their favorite fishing place in the mountains using GPS? Or that location-based services using GIS/GPS technologies would become widespread for a variety of applications, including taxis, mail, trucking, and other transportation services? The application of GIS/GPS and associated technologies does not stop there. Adding wireless technology into the mix provides two-way communication possibilities. There are other technologies, such as light ranging radar (LIDAR), digital sensors and instrumentation, high-resolution remotely sensed images, and laser ranging devices, that can be interconnected to GIS and GPS. Previously, hardcopy maps, charts, and notes may have guided us from one location to another. That is, we did not use the output in real time, updated moment by moment, and certainly not using a personal data assistant (PDA), as is now possible.

Spatial data refer to position, as compared to *aspatial data*, which describe attributes located at the position (Table 1.1). Geotechnologies not only allow us to identify and locate where we are and when we are there (spatial data); they provide the

Table 1.1   GPS-linked spatial and aspatial data

Object	Spatial Data		Aspatial Data		
	x-Coordinate	*y*-Coordinate	Height (ft)	Age (years)	Species
1	3	3	25	81	Pine
2	6	2	14	42	Fir
4	5	5	23	51	Spruce
5	7	6	14	33	Spruce
6	8	4	25	94	Pine
7	9	7	24	82	Fir
8	6	8	21	63	Aspen
9	5	9	18	67	Pine
10	4	1	23	71	Spruce

ability to assign attributes to a location quickly and easily (aspatial data). Spatial database tables, including associated attributes, can be constructed in digital format using GPS software. Aspatial data and attributes describing locations and entities can be monitored and recorded using GPS-linked instrumentation.

For example, assume that the goal is to monitor soil temperature over the course of a day. There are 10 locations in a field that will be monitored, and it has been decided to measure the temperature every 6 hours for 2 days. There are several ways that we can go about this. Using GPS, we first locate the 10 coordinates or *waypoints* and record them with GPS. This provides the spatial information. It will not be necessary to locate them more than once. Each site can be visited every 6 hours for 2 days. Recording the locations using GPS yields Table 1.2. This is a spatial database, indicating the positions where the soil temperatures will be taken. It should be noted that the waypoints are in meters. In this case the origin is located at (0,0), which means that each location is x meters and y meters from the origin. From this, soil temperatures are added as aspatial data. Since we talk about database structures and techniques later in the book, we keep this simple here.

One table will be constructed for all the data and will be similar to Table 1.3. To measure the soil temperatures, a thermometer is placed into the soil to a depth of 5.0 cm. A soil temperature measurement for each location is taken every 6 hours using this method. The temperature values are written on a piece of paper and are later input by hand into a spreadsheet, appending the spatial data table. The data table now consists of both spatial and aspatial data, with soil temperature values recorded every 6 hours. These spatial data, where the temperatures were taken, can now be loaded into a GIS. This is done by transferring the spreadsheet tables in ASCII format or a delimited text table into a GIS and plotting the locations (Figure 1.2).

The soil temperatures themselves can also be graphed from within the spreadsheet, as shown in Figure 1.3. The soil temperature trend of rising during the day can readily be seen. The temperature cools past midnight and rises once again in the morning of the next day. This table can be queried and thematic layers created, although for this data table, there might be a problem querying since there are two values for every time, duplicated each day. Thus when "1200" is queried, the user will

Table 1.2 (x,y) GPS coordinates

Location	x-Coordinate	y-Coordinate
1	538435.406	8076546.000
2	538487.719	8076549.500
3	538276.875	8076477.000
4	538332.438	8076484.500
5	538384.000	8076488.000
6	538439.063	8076491.000
7	538487.813	8076492.500
8	538276.531	8076427.500
9	538332.188	8076430.500
10	538385.844	8076433.500

Table 1.3 Soil temperature locations

			Soil Temperature Readings by Hours								
Location	x-Coordinate	y-Coordinate	0	600	1200	1800	2400	600	1200	1800	2400
1	538435.406	8076546.000	11	10	14	16	17	13	15	15	18
2	538487.719	8076549.500	14	12	17	19	20	16	18	19	17
3	538276.875	8076477.000	13	11	15	18	19	14	17	19	17
4	538332.438	8076484.500	14	10	16	19	20	13	16	20	16
5	538384.000	8076488.000	12	12	17	18	21	15	18	19	18
6	538439.063	8076491.000	15	13	15	16	18	12	17	18	16
7	538487.813	8076492.500	16	12	16	18	19	14	15	18	15
8	538276.531	8076427.500	15	14	15	17	19	13	16	19	14
9	538332.188	8076430.500	14	13	16	19	21	14	18	20	17
10	538385.844	8076433.500	13	15	18	20	20	12	17	19	16

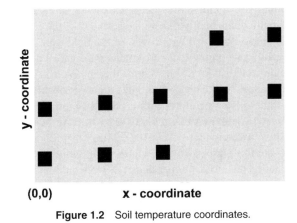

Figure 1.2   Soil temperature coordinates.

not know to which day it refers. A quick way to deal with this is simply to rename the times. We know that the study began today and runs for 2 days (48 hours), so the hours will be identified from 0 through 48. Alternatively, Julian calendar dates could be used, and data can be queried quickly.

Soil temperatures can vary over short distances due to a variety of factors, including organic matter, moisture, slope, time of day, instrument type used to measure the temperature, and instrument accuracy. This leads to the issue of *sampling intensity,* a measure of the number of measurements taken. In our example 10 soil temperatures are taken over a small area, 10,000 m^2 (1 hectare) here. Ten measurements are not very many; they were acquired fairly quickly and cheaply. This is one of the benefits of lower sample intensity—or is it?

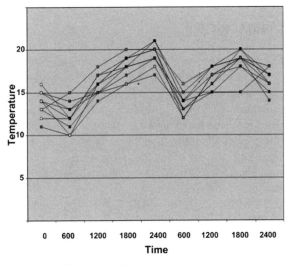

Figure 1.3   Plot of soil temperatures.

The 10 individual soil temperatures each represents an area of about 1000 m^2 (10,000/10). But that is not actually true, since a review of the figures shows clearly that some areas in the southeast and northwest of the field were not sampled for soil temperature. Given that we know that soil temperature varies across a field for the reasons already indicated, is it logical to assume that those areas without samples can be represented accurately by the nearest sampled areas? Probably not. Further, can it be assumed that the soil temperature within each 1000-m^2 area is homogeneous across the entire area? Again that is not likely to be true. Yet we have GPS position locations for 10 sites in a field that covers 10,000 m^2. The GPS locations were taken to identify and locate those areas from which the soil temperatures are taken. If we increase the number of positions (higher density), we would be increasing the number of soil temperatures taken from the field. The value of this is that it would more accurately reflect the temperature variation within the field. But how many are needed? That depends on the variability of the soil temperature in the field, the resolution needed, and once again, the cost. The higher number of samples taken requires more time for collection, processing, and analysis. They may, however, be more useful if the information is being used to drive a model that is constructed to predict and simulate soil temperatures for agricultural fields.

The period between which the samples are taken is referred to as the *sample frequency.* For the soil temperature study, the sample frequency is 8 hours. The *sample duration* is the duration for which the soil temperature data were obtained. In this case it was 2 days. Finally, once a soil temperature position is located, the sampling frequency and sampling duration can be increased for any given location without the need to reacquire the GPS position. In effect, it is relatively economical to increase sampling frequency and duration since all that is required is to visit more often and for longer periods of time, taking only soil temperatures, which can then be appended to the georeferenced locations in the data table, noting their times.

1.3 COMPASS READINGS

The compass remains useful for many applications in conjunction with geotechnologies. A compass is an instrument that includes a small magnetic needle that is capable of determining magnetic north. It is used to determine angular direction with respect to magnetic north and locating positions on Earth's surface. A compass may also be used to determine true north, provided that the proper angular offset is used to adjust the instrument. It should be noted that magnetic north and true north are at two different locations. *True north* is the definitive northern pole as determined through astronomic calculation. *Magnetic north,* which is near the North Pole, cannot be defined as an (x,y) coordinate since it is an area of magnetic influence that occurs naturally. Alternatively, true north can be defined by coordinate. The difference between magnetic north and true north is often referred to as *magnetic declination.*

In many geotechnology applications involving GPSs, a handheld navigator may not be able to acquire a satellite lock on available satellites, effectively rendering the

GPS navigator useless for recording waypoints. This can happen under dense forest canopies, within or near tall buildings, when it is very wet, and when terrain conditions are such that taller mountains and obstacles limit the line of sight to GPS satellites from the GPS navigator's position. Another important matter to consider when using electronic geotechnologies is why a compass is often useful. Geotechnologies tend to use significant amounts of power and at times may be susceptible to breakage or electronic malfunctioning. Either loss of power or inoperable equipment can lead to an inability to capture positions and record them. A compass provides the means to determine angles and together with walking or pacing can become one means of locating and distinguishing one position from the next. In such cases, field notes should be kept that indicate the last known GPS coordinate, carefully noting other locations from that point when using a compass. This information recorded manually can later be integrated into the downloaded GPS spatial data tables and imported into a GIS.

The offset declination to use for a compass will vary depending on the user's location. Declination tables are published annually for locations around the globe, and a compass should be adjusted for use based on those tables. If magnetic north rather than true north is needed, no offset declination is required and the data can be transformed later. There are several other reasons to use a compass when doing GPS work in distant or remote locations. First, in the event of equipment malfunction in a remote area, the user has no other way to determine position and can become lost. Second, GPSs are not magnetic but electronic and without moving, cannot determine an azimuth. If one is in the unfortunate state of being unable to move, one cannot determine direction with a GPS—at best, several waypoints will be presented surrounding the current position.

1.4   NAVIGATION AND AGRICULTURE APPLICATIONS

GISs and GPSs are useful not only for locating positions on Earth, but have become useful for navigation and location-based services (LBSs). Navigation may or may not include real-time applications. In real-time applications, the GPS position is updated continuously. GPSs may be used to track animals, automobiles, airplanes, boats, bikes, and other objects. We might want to know where a car is in a city at any given time, where an international airline flight is, where a train is, and the locations of health-related concerns in our communities. We may want to be able to locate a house, plan a trip across town or to the country, or determine an optimum route as to travel time and estimated time of arrival.

You might want to know how fast a GPS satellite is moving in space. That calculation will put into perspective how truly amazing GPS satellite technology is, particularly when considering navigation applications on Earth. To determine this, a bit of leeway regarding distances has been taken. To begin, let's assume that a GPS satellite is about 20,000 km distant from Earth. It then takes the satellite about 163,000 km to go around the Earth, so the satellite circles the Earth twice daily, thus covering about 326,000 km in the course of a day. Dividing that distance by 24 hours yields a travel speed of about 13,500 km per hour! It is difficult to imagine calculating ground accuracy within a few centimeters from satellites moving at that speed—but that is essentially what happens.

The ability to navigate more accurately using GPS/GIS technologies can have economic benefits for many applications. Precision agriculture couples GPS and GIS technologies. For this, agricultural fields are mapped with GPS and a contour map can be generated from the data table in GIS and a digital elevation model (DEM) constructed for the field, providing continuous surface (x,y,z) coordinates (see Figure 1.4). The z coordinate is the elevation in meters. A 10-m contour interval was used in this figure, showing that we wanted to generate index lines or contours where there is a change of 10 m in vertical elevation across the field. Since only 20 original values were used in the table, the remaining elevations for the field had to be generated by interpolating or estimating.

In precision agricultural applications, many people use GPS/GIS to manage inputs, increase efficiency, and reduce environmental problems. At the same time, land use policies are being considered which could result in much wider adoption of precision farming geotechnology. The Nitrogen and Environmentally Sensitive Area Scheme and the EU Nitrification Directive in Europe are designed to consider environmental effects on farming. Agricultural policy in the near future may shift from conventional farming techniques to a greater use of geotechnology (Blackmore, 2000). Geotechnology would provide monitoring and agricultural management/operational capability, ensuring that such policies are met. In this scenario, each agricultural producer would maintain geotechnology data.

The elevation contours that we have generated form the basis on which precision agricultural operations may evolve. In the case of herbicide applications for agricultural lands, a producer will survey agricultural fields to ascertain any weeds present, then select appropriate herbicides for application. In some cases this may mean more then one herbicide application, since herbicides tend to be selective, affecting only specific species of weeds. Most weed populations are not distributed evenly throughout a field. Instead, they are found in patches where larger infestations occur. It would be far more efficient to locate only those areas that are infested with weeds and to

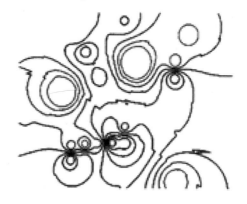

Figure 1.4 Contour map.

spray those while decreasing herbicide costs. This approach achieves necessary weed control while reducing adverse and potential environmental problems. In practice, a producer would travel over the field, adding to a map the species and locations of weeds (Figure 1.5). Since that map is derived from an existing map of the field, it can be used as a thematic layer together with a DEM to show weed locations. To do this the weed areas drawn will need to be digitized, which can be accomplished on-screen or by using a digitizing tablet. The weed theme is then saved in the GIS; alternatively, the map could be scanned, then incorporated into a GIS, and using previous GPS locations used for constructing the DEM, it can be georeferenced.

Human beings are highly skilled at classification and pattern recognition, particularly when static, two-dimensional entities are human-made or capable of being examined in isolation (Burrough and Frank, 1995). There are cases, though, where classification and pattern recognition, such as the earlier weed example, do not have clear, definitive boundaries. These are called *indeterminate boundaries*. Weed infestations tend to be in patches having irregular patterns lacking sharp boundaries. Although polygons can be mapped with a GPS, where exactly does one draw the line or boundary? One alternative is to use a weed infestation rating system. By rating weed infestations based on their level of infestation (i.e., on a scale ranging from 1 to 10), 10 classes are established. The location of the weed areas provides spatial information, and the classes provide attributes and values or descriptors of those spatial positions— becoming aspatial data.

Developing such classes has much practical value, particularly since one class can be assessed against another class once imported into a GIS. Alternatively, two or more themes can be analyzed together to gain a better understanding of the overall weed infestation level of a field. Many questions using GIS can be answered with respect to position. Where and what total area are covered in weed infestation class 7? Are the infestations distributed evenly throughout the field or dispersed more widely? Are most of the ratings above 5 or below 5? By answering these questions, planning can begin toward the development of a spray program using GPS technologies coupled to agricultural field equipment used for herbicide application. In operation, those areas requiring herbicide would be marked (i.e., all areas of a class above 5) and their coordinates uploaded to the tractor GPS. Coordinates would trigger the turning on or

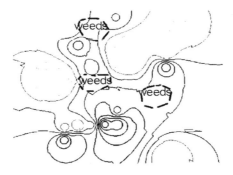

Figure 1.5   Weed locations on contours.

off of the herbicide sprayer as it moves across a field. Weed infestation is only one example, but these techniques and technologies can be applied to almost any situation where indeterminate boundaries exist.

There are other geotechnologies that can be considered in this example. Aerial photographs or satellite imagery could be used for mapping weed populations. In the case of satellite imagery, the levels would be determined through the use of image analysis, interrogating pixel values, different pixel colors providing an indication of weed infestations. Alternatively, photogrammetry analysis for texture, tone, and shape may reveal weed areas. In each case the boundaries of the weed infestations would have to be delineated, then a georeferenced thematic layer created and subsequently uploaded to field GPS equipment for final herbicide application. Since economic advantages are the goal of precision farming, only those areas identified would be sprayed using this method, and the producer can select from a range of levels which are to be sprayed based on the economics, the desired yield, and the level of weed control to be maintained. Applications of a similar nature include crop dusting, fertilization, road and highway chemical applications, maintenance applications for fields, and applications where position is necessary to trigger events or operations.

1.5 TIME AND DISTANCE DATA

All spatial events will have a time component. Some may be static, seldom or never changing, whereas others may be dynamic, changing constantly. This applies to the measurement of stream temperatures in the course of a day, which is dynamically changing or may include moving vehicles within an hour. It could also include aerosols surrounding road networks during peak travel times in cities. The physiological stages of the growth of a wheat plant over a season to the altitude of an airplane during a flight are also considered dynamic changes. The movement of a swing crane on a construction site and the operation of snow removal equipment in a city also include a time element. The telecommunications connections between banking machines depend on time. Time is used to synchronize the transfer of digital data packets. The time base used for automated banking machines (ATMs) is a GPS. From the moment a pizza is ordered from a delivery dispatcher to the selection of the nearest outlet and subsequent delivery to the customer, time is involved. In such a case, the nearest outlet (distance) and route planning (time) are evaluated using a GIS.

The connection between telecommunications and spatial databases, termed *telegeoprocessing,* is a new discipline based on real-time spatial databases updated regularly by means of telecommunication systems and used to support online decision making. The economical (and useful) production of aerial photographs for stereoscopic overlap use depends on time. The path of an aircraft or flight line and their relationship to each other are determined by using GPS, thereby avoiding unnecessary side or end lap, the amount of overlap of two aerial photographs. The amount of overlap is a function of time—the speed of the aircraft in relation to shutter opening and focal length. In this case GIS and GPS technologies together are capable of acquiring time, monitoring time, and analyzing events with respect to time and the flight path.

Why has time suddenly become so important? It is through observing locations (entities) and events through time more closely that we are able to understand their character and nature, the premise being, of course, that if we understand their character and gain knowledge about them, we will be able to manage them more effectively, efficiently, and with higher levels of quality. There are also economic advantages to improved location and navigation of events through time. GIS and GPS enable these scenarios, noting that both technologies include location and navigation as well as providing the ability to analyze them. Not only can we determine where we are moving in real time, but we are also interested in knowing how one event relates to other events happening at different times. How does an event at time A compare to a similar event at time B? The U.S. Geological Survey (USGS), for example, monitors stream-flow conditions in real time nationally, providing a comparison between streams based on flow rates. This is an example of data logging/instrumentation technologies coupled to GIS. Are the attitudes of one region this year similar to these five years ago, or have they changed? What are the demographics of the region now as compared to five years ago? What influences buyer attitudes in the region, and are they true for the entire region or merely for a few blocks? How did we gather the spatial data to make that judgment, anyhow? GPS is used to locate people, their homes, and if they live in a high-rise, the floor they live on. This is often referred to as *geocoding*. When coupled to other demographic information (e.g., census data), this positional information serves to provide clues and answers to these and other questions. But it is through the integration of GIS and GPS that these questions can be addressed with higher degrees of accuracy.

Another aspect of time is related to historical events. Some people are interested in real-time applications; others, such as archaeologists, modelers, and historians, are interested in locating past events. They seek information about where and when events occurred. In the case of global climate change, many institutions and individuals are attempting to model past weather events. The value of this is to provide an estimate of current model accuracy based on the ability to replicate past events and whether or not models can be used to predict future events accurately. This is not an easy task given that the world population (a stressor) changes over time and that the rate of industrial expansion (another stressor) is changing while land use changes dynamically and does not remain static (an additional stressor).

Aerial photography has a long history and images can often be found dating back a decade or longer. Many national parks and government agencies maintain large collections of historical images, as do many cities. Historically, the surveying community has been largely responsible for landscape measurement. Land survey records and deeds created by surveyors exist in many historical archives and offices, as do survey records for locating railways, waterways, transportation, and cities along with individual property titles. Cadastral mapping has a long history that remains important and will continue to do so well into the future, providing a means of physically measuring and determining land division. Much of that information remains to be placed in a digital format that requires data conversion. After all, it is the comparison of historical information to current information that will provide the identification of change over time.

Monitoring time as it changes reveals other issues for a GPS/GIS professional. *Network analysis* involves locating entities and events and analyzing them with reference to a network. A road system in a city is a network, as is a collection of computers or an airline, rail, or delivery system. These can be modeled using *allocation–reallocation models,* which might be used to determine both time and distance with respect to volumes of traffic, population, and time of day. An analysis of networks involves both time and distance, although many people consider distance to be the primary factor. That is not necessarily so. Many of us dream up ways of getting to work quickly and home even more quickly (although a few like work so much that they are seemingly always there!). Or perhaps they take work home, thereby shortening the distance and perhaps expanding usable time. Assuming that most of us go to a place of work, then return home, knowledge of the road network is very important. Traveling time from home to work may require crossing a city. For a network to operate there must be *connectivity;* roads must be connected so that traffic can flow from one road to another smoothly and in a continuous fashion. In the example illustrated in Figure 1.6 we assume that the road network is highly connected. For this network, home is at point A and the workplace at point B. The travel distance between them is about 1 km. As can be seen, there are numerous routes available for traveling from A to B, but the one shown with arrows is the shortest-path distance, as determined by network analysis using a GIS. Network analysis determines the two points and assesses the connectivity of the road network to determine the optimal path that would provide the shortest distance.

There are many benefits in knowing the shortest route between two points; however, a distance-only analysis can provide only a general sense of time. Distance-determined routes may be the shortest, but with respect to the time, may be the longest. Some of the factors that influence the time required to travel the path include:

- Current traffic volumes
- Weather
- Topography
- Accidents
- Height and width restrictions

Although other routes are longer, they may be quicker if some of these conditions are more favorable on those routes. Therein lies one of the more interesting issues involving network analysis—time. Both position and time must be considered when evaluating a network. Using GPS in our earlier example, elevations were recorded along all roads of the network. These elevations were then used to generate a *triangulated irregular network* (TIN) in GIS software. A TIN is a computer algorithm that connects individual elevation points into a series of triangles, thus forming a continuous surface (Figure 1.7; see Chapter 4). As the arrows indicate, the shortest-path distance crosses some of the highest points on the terrain. It is likely that larger trucks would move up higher elevations more slowly, and the possibility of cars climbing steeper slopes could be problematic under wet or snowy conditions, therefore con-

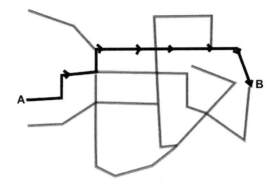

Figure 1.6 Connected road network.

tributing to slower speeds. Alternatively, routes to the south have lower elevations, and although these routes may be longer, they may have quicker-moving traffic. Thus, a distinction is again made between travel time and travel distance.

1.6 INTEGRATIVE MODELING

Spatial modeling also involves time. A *spatial analytic model* is a set of numerical expressions implemented in stand-alone software principally for describing, explaining, and possibly predicting a particular, special aspect of past, present, or future (imagined) georeality (Nyerges, 1992). These models are used for determining such things as weather patterns through time and into the future, growth of forests, and the changing status of the world's oceans. There are a growing number of individuals and organizations interested in modeling these events—coupling GISs to models. For a model to operate requires numerous types of data (depending on the complexity of the model) and often that data be collected and analyzed with a GIS or GPS prior to being input into powerful supercomputers where the model is run.

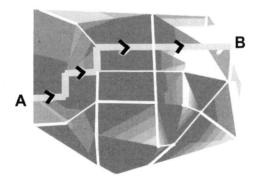

Figure 1.7 Road network and TIN.

Not all models require a supercomputer, only the more complex ones. The *accuracy* of a model—its ability to re-create reality accurately—depends largely on the quality of the information included in the model. That is not an easy task since many phenomena are not static but both change and involve anomalies. For example, the estimation of leaf area index from forests has been shown to change during the course of a season, as would be expected, but estimation is compounded by the variation in spectral reflectance using advanced very high resolution radar (AVHRR) (Spanner et al., 1990). A GIS or GPS is often used to collect, analyze, and collate information prior to its incorporation in a model. Generally, output accuracy is improved when higher-resolution data are used. For this, a GPS can be used to collect very detailed local information, locating entities with both higher accuracy and higher precision, while a GIS imports and uses GPS data in an analytical manner, tailoring the spatial data for final input into a model. Sensor and instrumentation technologies play a very major role in acquiring data for agents that drive a model. More recently, remotely sensed information has been proving very useful for models with resolutions similar to those for GPSs.

Modeling is gaining in popularity due to the fact that solutions may be formulated and compared in both an exploratory and applied manner without the need to actually collect large quantities of data. Financial savings are one of the benefits of using modeling, and they also attempt to provide solutions to real-world questions. Artificial intelligence is already being linked to GIS applications using artificial neural networks, fuzzy logic, and evolutionary computation, among other artificial intelligence algorithms. The ability to model events and their interaction and outcomes directly involves time and scale.

1.7   SCALE AND GENERALIZATION

The scale or representation of geographic information provides the viewer with an opportunity to see more or less detail. *Large-scale representation* refers to a closer view, allowing for observation of more detail. *Small-scale representation* involves the observation of information from a more distant vantage point. For example, a small-scale representation may include an area covering several hundred square hectares, whereas a large-scale view may only represent a river, creek, or roadway, with corresponding higher detail. A GPS is able to acquire spatial data regardless of scale. If the user wishes to sample every 10 m (enabling large-scale representation) or every 500 m (enabling small-scale representation), each can be accomplished. Whereas the data from 10-m intervals may be integrated to provide 500-m interval data, the 500-m data cannot be reduced to 10 m without interpolation. Thus, data can readily be scaled up but not as readily scaled down, at least not without significant interpolation that calculates intermediary values between sample points. This poses interesting questions for geotechnology users because so many geodata from a multitude of sources are not standardized to a similar scale.

Much of the challenge of modeling interactions between natural and social processes has to do with the fact that the processes in these systems result in complex temporal–spatial behavior (Itami, 1994). Ecological models are notorious for their complexity, notably the high numbers of agents and stressors and the complex inter-action between them that collectively form a system. High variability in vegetation population, slope considerations, climate, plant physiological stages, and fluctuating mortality rates, for example, all contribute to the complexity and difficulty of their process modeling. These variables are highly nonlinear through space and time, which has important implications where geotechnology is being used. Widely spaced GPS waypoints along a stream, which are assigned values associated with flow rate, depth, and temperature, may not adequately address the scale at which these processes occur and are presented. Even the physical location and dimensions of the stream may not be established correctly (e.g., transformation errors).

Generalization refers to the level of visual detail in a representation. There are no assigned values for generalization, nor is there a scale indicating amounts of gener-alization. Instead, generalization is expressed as a comparative measure between two or more representations. For a stream, a map product with high generalization would show less detail, such as sharp bends and variable physical dimensions. A map product with low generalization would show more detail and provide a more visually realistic representation of the stream, including sharp bends and more accurate physical dimen-sions. It is important to realize that a map, either hardcopy or on a computer screen, can have lower amounts of generalization and yet be inaccurate (e.g., the data are of poor quality). Sharpness and clarity are not necessarily synonymous with quality.

Let's assume that there is a map for a mountainous area near Oslo, Norway, at a scale of 1 : 100,000. GPS waypoints have been collected using a regularly spaced grid of 50 m to develop a 1 : 100,000 map for a park area where tourists will visit and spend the day. The GPS data are very accurate, and some interesting landmarks were noted while collecting the data. The GPS staff was quite proud to have achieved less than 1-m accuracy during the GPS survey. Later, tourists are handed a map with locations of historical sites and may choose to go wherever they wish. Six hours later a few tourists arrive back at the embarking point, tired and grumpy, and state: "We could not find the historical sites." Their clothes are tattered and the tour guide does not understand the problem: They had a good map prepared using excellent GPS data, so what's the problem?

Even though the GPS data collected were highly accurate, the maps generated had a scale of 1 : 100,000. Subsequently, many of the details of the landscape were omit-ted (i.e., the features were generalized). No wonder the tourists were angry and upset—they had run into small creeks and wet areas that did not show on the map. Luckily, though, they did find paths to get back that also were not on the map. This was not due to the fact that the maps were old and the paths new but that the generalized map did not show features in detail. The guide was fortunate that the tourists found the pathways; otherwise, the situation could have become very tense.

For this reason it is often said that data collected at one scale should ideally be rep-resented at a similar scale. One should attempt to reduce generalization as much as possible and remain aware of it. There are economic considerations concerning gen-eralization. There is not much point in collecting detailed information at a higher cost when it is not going to be represented or used. In fact, it could be argued that use of an alternative technology might provide more economically beneficial for collecting data to be represented at one scale as compared to another. Individual geotechnologies are capable of varying levels of accuracy, so are designed and more useful for particular types of applications.

1.8   VISUAL COMMUNICATION

Visualization has become increasingly important as related to applications involving GPSs and GISs. The value of visualization is in its ability to be used for communication and exploration of spatial information. One researcher (MacEachren, 1994) has noted that visualization is useful when delineating the known from the unknown while moving from a private to a more open or public experience (Figure 1.8). Another researcher speaks about the ontology of perception, describing how attributes can be perceived and represented alternatively in topological structures (Gahegan, 1998). For their reason, many organizations are becoming interested in visualization of spatial information for this reason.

Visualization provides a means to communicate to the public information about research, future real estate developments, environmental problems, population statistics, political boundaries, recreational proposals, fishing habitats, and climatology, to name a few. Geotechnologies provide a means to collect, manage, and analyze spatial information that is later visualized in two, three, or four dimensions. GPS is very useful for capturing spatial and aspatial information to be visualized, due to its ability to record events with time. This permits visualizations to be created that are dynamic and change through time. Coupling visualization to the robust analytical capabilities of a GIS, however, remains limited (Germs et al., 1999). Visualization does not necessarily mean that a connection to GIS is present, although there would be benefits to

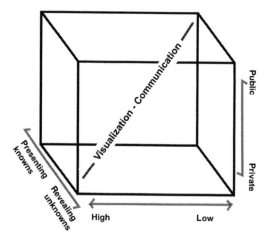

Figure 1.8   Visualization and communication. (After MacEachren, 1994; with permission from Elsevier Science.)

a closer coupling of visualization to GIS (Thurston et al., 2001). Visualization attempts to mimic the camera, whereas GIS is more closely related to the database (Figure 1.9). This results in a division between GIS/GPS technologies and visualization.

GIS can query data tables, rendering static thematic layers, which are then transferred to visualization software, where traditional image enhancement and presentation can occur. The question becomes: How can a GIS database be queried in real time and the query presented with a high degree of image enhancement using visualization and graphics tools? Let's take a closer look at the integration of GPS/GIS and visualization. *Draping* is one method of producing a photo-realistic image on a GIS DEM. First, a digital terrain model (DTM) differs from a digital elevation model (DEM). A DTM is a model that represents the Earth's surface for a given region in three dimensions (3D). A DEM is a model that represents all objects, including the surface for a given region in 3D. It is easier to create a DEM than a DTM (both of which are actually 2.5D in GIS; see Chapter 4). For a DEM all that is needed is LIDAR instrumentation and an airplane. This will provide a model of the surface and objects on the surface, noting their height (elevation). A DTM, by contrast, requires that the Earth's surface be represented accurately beneath trees and other objects, something that LIDAR may not be capable of achieving, whereas a ground survey is capable of accurate representation. Many people use DEM and DTM interchangeably, but they are different. If I were to order a DEM, I would expect to receive a continuous surface with buildings, towers, cars, boats, and other objects shown in 3D atop the Earth's surface, also represented in 3D. But if I ordered a DTM, I would expect to receive a model showing only the Earth's landforms in 3D.

Both DEM and DTM can provide a continuous surface. A 3D continuous surface can be produced in a GIS using a triangulated irregular network (TIN). To produce a continuous surface, a series of (x,y,z) coordinates are taken across the landscape using GPS for the area to be visualized. The GPS waypoints can be taken 1 m apart, 100 m apart, or at any other grid spacing that is desired. Selection of the spacing will directly affect the accuracy of the DEM produced. Grid spacing with coarser, more widely spaced points will result in few locations having (x,y,z) coordinates, which then require more interpolation for areas not sampled. Using finer grid spacing or less dis-

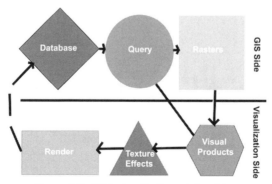

Figure 1.9 GIS–visualization relationship.

tance between GPS waypoints to acquire the (x,y,z) coordinates will result in more data, less interpolation, and subsequently, more accurate representation of the topography of the area. There is a trade-off when determining grid intervals. Collecting more points (finer) will require more time and thus be more expensive. Too few data points collected will be cheaper, may not accurately reflect the terrain being mapped, and will be less useful. Due to the time and higher cost of building a 1-m vertical DEM, they are not used very often, particularly for larger areas. More commonly, DTMs with 10- to 100-m vertical elevation changes are used. This coarser resolution affects the quality of visualizations that are created for landscapes. Smaller grid spacing would be more useful for visualizations representing near photo-realistic quality. This would ensure that smaller elevation changes (e.g., less than 10 m) more accurately represent the terrain being modeled and visualized. Smaller hills and ridges would be captured with finer detail, enabling truer representation. For general visualization a coarser grid may be adequate, but for scientific research, cinematography, and applications requiring finer detail, more closely spaced gridding is necessary. Finer grids are also more useful where emergency applications are involved.

GPS is a primary data capture tool used in landscape-based studies leading to visualization. It can be used to capture spatial information and attributes about positions at the same time. This is important when scenes that are to be rendered later include objects that must be located accurately in time. Since GPS can usually acquire a position within ± 10 m and usually much less, objects to be visualized can be positioned with a high degree of accuracy. Assume that there is an open traverse running from point A to point B in Figure 1.10. GPS has been used to collect the (x,y,z) coordinates for the points to 1-m accuracy or less for the purposes of later placing a road along the points using visualization software, represented as circles. The traverse is placed atop a DEM that has a grid spacing of 100 m, meaning that the DEM represents elevation changes that depend on values using that grid spacing. The DEM elevations are represented in 3D as evenly spaced bars and their heights are provided in the key.

This is an interesting exercise because it involves the integration of DEM, GIS, and GPS. A review of the DEM shows that in the first row of the grid, elevation is about 10 to 30 m. In the second row of the grid where the proposed road is going to

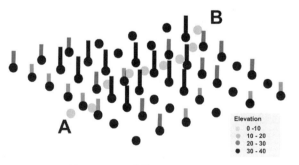

Figure 1.10   DEM and traverse.

go, the elevation is 80 to 110 m, then decreases slowly moving through the grid. There is no way of knowing if the first row of the grid rises evenly from 10 to 110 m toward the second row. Is it 110 m for most of the distance from point B to point C, or mostly 10 m in elevation, rising sharply to 110 m at point C? If the first and second rows of the grid were interpolated, we might find the average between the two points but still not be sure that is accurate since the terrain between points B and C could undulate significantly, perhaps even dropping below zero sharply, then rising again.

The question often asked is: What is the point of collecting 1-m GPS information if the representation grid is at 100 m? The GPS information and existing GIS values for attributes can be integrated to a smaller grid. In this case the grid could be changed to that representative of the GPS, or 1 m. In this way the z values as recorded by the GPS would become part of the data set and could be used as values between the first and second rows of the grid—essentially increasing the number of values for elevation in that region and subsequently, the accuracy. This would then mean that areas visualized around that region are much more accurate (reality), leading to a higher level of representation. It also means that any 3D objects in addition to the road itself could be rendered more accurately with respect to terrain. Again, smaller grid spacing equals higher reality.

The bars would continue to be represented at 100-m spaces, and each new value for the spacing between existing rows would have an interpolated elevation value. An irregular grid spacing can be used to build the DEM. Let's assume that we use the same number of points as were used in the evenly spaced grid but instead take more GPS waypoints (x,y,z) nearer the area where the road is to be visualized (Figure 1.11). The concentration of GPS waypoints taken near the path of the proposed road provides a more accurate representation of the terrain for that area because the waypoints are based on real elevations that have been collected, with fewer interpolated values. The trade-off in achieving a higher representation for the proposed road area is that the more distant areas, such as those on the right-hand side of the figure, are not sampled as frequently. Subsequently, those areas will have less accuracy and require more interpolation if a DEM is to be created. Both regular and irregularly spaced samplings can be termed a *model,* each method attempting to model or represent the visualized area under study.

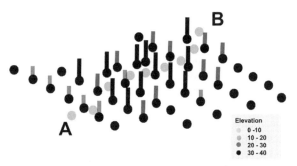

Figure 1.11   Irregular GPS grid sampling.

Reviewing both the regular and irregularly spaced grids, it is clear that evenly spaced GPS waypoints will provide a general level of accuracy for the DEM throughout the entire area. The irregular spaced grid will provide a higher level of accuracy for a smaller area at the expense of less accurately representing areas more widely spaced. The selection of waypoint spacing for building DEMS directly affects the ability to render the scene accurately in both height and scale. In choosing which spacing method to use, consideration of accuracy and scale must therefore be considered. Three-dimensional objects can be created using 3D modeling software. Objects can include trees, houses, people, cars, ships, rocks, or other objects. Accurate placement of these objects into a scene depends on knowing their true locations. GPS is ideal for locating object positions. Surfaces in visualization may include landscapes as DTM or DEM or even a room in a building that has been derived from computer-assisted architectural drawings (CADs). In each case, (x,y,z) coordinates are used so that the objects can be attached to vector (or raster) coordinates on the surface. We discuss the merits of raster and vector representation more closely in Chapter 6.

1.9   DATA STANDARDS

In 1994, the U.S. president indicated:

> Geographic information is critical to promote economic development, improve our stewardship of natural resources and to protect the environment. Modern technology now permits improved acquisition, distribution and utilization of geographic data and mapping. The National Performance Review has recommended that the Executive Branch develop, in cooperation with state, local and tribal governments and the private sector, a coordinated National Spatial Data Infrastructure (NSDI) to support public and private sector applications of geospatial data in such areas as transportation, community development, agriculture, emergency response, environmental management and information technology. (Executive Office of the President, 1994)

Several countries—Japan, New Zealand, the United Kingdom, France, and Sweden, among others—now have a national spatial data infrastructure in place and are developing new internal standards. In Canada, the Canadian Geo-spatial Data Infrastructure (CGDI), through Natural Resources Canada, handles national standards and the development of integrated spatial data sharing. Many countries are interested in and currently developing international geospatial data standards, but in many cases these are not yet in place. In some countries there are issues related to the ownership of data. In many cases, agencies do not wish to freely distribute their geospatial data. In private industry, geospatial data are largely considered to be intellectual property, with a value attached to the data. Often, it is not desirable that this proprietary information be made available to competitors.

The Geomatics Industry Association of Canada (GIAC) defines *geomatics* as "a technology and service sector focusing on the acquisition, storage, analysis, dissemination and management of geographically referenced information for improved decision-making." This is to include surveying, mapping, remotely sensed products and services, together with global positioning systems (GPSs) and geographical information systems (GISs). Geomatics can also be defined as those information technologies that are used specifically for geographical purposes.

The Canadian Geo-Spatial Data Infrastructure (CGDI), a national clearinghouse operated by the Canadian government, designed to integrate and disseminate geospatial information, states: "Geomatics is the science and technology dealing with the character and structure of spatial information, its methods of capture, organization, classification, qualification, analysis, management, display and dissemination, as well as the infrastructure necessary for the optimal use of this information."

The Canadian Institute of Geomatics (CIG) says:

Geomatics is a field of activities which, using a systemic approach, integrates all the means used to acquire and manage spatial data required as part of scientific, administrative, legal and technical operations involved in the process of the production and management of spatial information. These activities include, but are not limited to, cartography, control surveying, engineering surveying, geodesy, hydrography, land information management, land surveying, mining surveying, photogrammetry and remote sensing.

Internationally, the Global Spatial Data Infrastructure (GSDI) states:

The Global Spatial Data Infrastructure supports ready global access to geographic information. This is achieved through the coordinated actions of nations and organizations that promote awareness and implementation of complementary policies, common standards and effective mechanisms for the development and availability of interoperable digital geographic data and technologies to support decision making at all scales for multiple purposes. These actions encompass the policies, organizational remits, data, technologies, standards, delivery mechanisms, and financial and human resources necessary to ensure that those working at the global and regional scale are not impeded in meeting their objectives. (GSDI, n.d.)

As more definitions are added to the list, there is a focus on georeferenced data that are captured, manipulated, analyzed, and output. The list grows and changes regularly as new technologies are developed and/or new applications for current geotechnologies are created for geospatial purposes. *Geomatic quality* is defined as "the accurate organization, location and display of geographic information." The International Standards Organization (ISO) working group (ISO/TC 211 SWG-QC) is designed specifically to deal with quality issues related to geomatics data, and ISO/TC 211/WG 1 is the specific ISO information body responsible for geomatics. Since the field of geomatics is relatively new, numerous definitions and proposed standards are continually being developed.

Issues related to the sharing of information between individuals and countries using common protocols and data networks related specifically to the Internet are slowly being resolved. One group, the Open GIS Consortium (OGC), an international group of institutions and companies concerned with seamless data sharing and ease of transfer, is working toward one common interoperable system. Under OGC, spatial data move freely between computing and technology systems for those who capture information and those who disseminate it. Such a system would be beneficial where different types of equipment and protocols generate results in numerous formats. OGC proposes: "OpenGIS is defined as transparent access to heterogeneous geodata and geoprocessing resources in a networked environment. The goal of the OpenGIS Project is to provide a comprehensive suite of open interface specifications that enable developers to write interoperating components that provide these capabilities...."

The statements and definitions above indicate clearly that we are not only interested in GPS and GIS when discussing geospatial data, but are increasingly aware that other forms of geospatial data can be included. Therefore, geomatics includes numerous sources of information collected from differing geographical locations around the world through time. Collectively, these technologies measure or record spatially based phenomena and their associated changes. GPS is one geotechnology being used increasingly for spatial applications. Currently, a U.S. and Russian GPS satellite systems exist. There are ongoing discussions in Europe on the development of a new European-based GPS system. This system, named Galileo, is supported by numerous European governments, which approved the project in 2002. The European Commission, Directorate for Energy and Transport, proposes a constellation of 30 satellites. High-resolution DEM information from the Shuttle Radar Topography Mission (SRTM) in the United States is also becoming available, with some restrictions. This information will be very useful for many geotechnology applications, provided that policies can be reached that allow general public use of the information.

Geotechnologies include, but are not limited to, GPS and GIS. Other technologies have a role as new applications are developed for small- and large-scale projects that involve time and space. Applications will vary and some may include modeling and simulation. Others will incorporate visualization as a means to express and explore new possibilities. Cartography will remain the backbone as these applications are developed because ultimately, spatial and aspatial data will be represented. If not represented properly, even the most accurate and complete data sets can lead to misunderstanding and confusion. Increasingly, there is a need to understand how geotechnologies relate to each other and how they can be applied in an integrated fashion. Understanding this integration requires that technology not be the only focus but that sound theoretical knowledge about the processes being mapped and represented be present. Ultimately, we are interested in answering where, what, when, how, and why and understanding the processes related to daily living.

EXERCISES

1.1. Define *geomatics* briefly.

1.2. Add 10 more locations to Table 1.1 in a spreadsheet. Transfer them to a GIS, make a map of all 20 locations, and provide a legend by species.

1.3. Make a data table using a spreadsheet that begins on a flat terrain rising to a mountain, then descends to a flat terrain. Use an evenly spaced grid of 300 m. Assuming that the terrain is a DEM, what sorts of applications could you use it for?

1.4. What is *enterprise technology,* and how does it relate to organizations with respect to GIS/GPS?

1.5. Describe two types of models and their advantages and disadvantages.

1.6. Spatial information may be collected in either two or three dimensions. When comparing them, discuss technological considerations for each and any other considerations.

1.7. Compare spatial data with aspatial data. In table format, provide information about a road network, indicating spatial and aspatial data.

1.8. *Generalization* refers to the ability to visualize varying levels of detail. What causes a product with a low level of generalization to be inaccurate?

1.9. Are large-scale aerial photos more or less generalized than small-scale aerial photos? Explain.

1.10. What is the difference between a DTM and a DEM?

1.11. One method of sampling elevation utilizes an evenly spaced grid; alternatively, an irregular grid may be used. What are the advantages and disadvantages of each method?

1.12. Why are standards for communication between computers necessary? What is *interoperability?*

2

GEODETICS

2.1 INTRODUCTION

Study of the Earth's physical shape is called *geodesy*. The shape of the Earth affects how geospatial information is collected and represented. The Earth is wider when measured around the equator than when measured around the poles. That difference is almost 67 km and complicates matters when attempting to represent three-dimensional space on a two-dimensional plane, as when mapping. Pythagoras, a Greek mathematician and philosopher (ca. 580–500 B.C.), determined that the Earth was a sphere. In 1687, Isaac Newton estimated that the Earth is not a sphere but in fact is geoid in shape, affected by gravity, thus resulting in an ellipsoid. The Greek astronomer and mathematician Ptolemy (ca. A.D. 90–168) provided a map system based on coordinate geometry, and going one step further, added a north arrow to the map. This had tremendous advantages, of course, readily allowing any user of a map to orient position with respect to a common direction. Ptolemy developed a system wherein a circle was based on 360° with each degree having 60′ (minutes). Each minute was then further divided into 60″ (seconds). This system constituted the basis for longitude and latitude (Snyder, 1993).

Accurate positions could not be located accurately in Ptolemy's time because of the coarseness of his calculations. However, this led to further understanding as to why the geoid shape of the Earth has such a large impact on determining the accurate location of a position on the Earth when using GPS and GIS technologies. In 1569, Gerardus Mercator proposed the *Mercator projection* (Snyder, 1982) for navigational purposes, still one of the most widely used projections. When spatial data are collected, reference to the shape of the Earth should be collected at the time of capturing the data. For example, a 2D GPS waypoint (x,y) may use the North American Datum 1983 (NAD83), World Geodetic Datum 1984 (WGS84), or another datum that more closely approximates the Earth's shape where the GPS waypoint is being collected. Like surveying, GPS data collection can be used to collect and measure lines, points, and angles on the Earth's surface. More specifically, this involves:

- Locating and measuring horizontal positions of the Earth's surface
- Locating and measuring elevations
- Locating and measuring physical shape and characteristics
- Points, lines, and polygons (i.e., directions)
- Locating and measuring line lengths
- Locating and measuring boundaries

Various types of surveys include the use of GPS. *Control surveys* are used to determine vertical and horizontal positions on the landscape, which depend on national geodetic networks (Craymer et al., 2001). The horizontal and vertical positions are then tied into known survey monuments or benchmarks that have been located precisely. It is not uncommon to find traditional survey equipment coupled with GPS equipment, particularly where the GPS is used to determine benchmarks. The determination of lines, along with their distance and direction with respect to boundary areas, is called *cadastral* or *boundary surveying*. An example of this type of survey would be in determining blocks and lots within a city. *Topographic surveys* involve the determination of elevations, which are useful for view-shed analysis to determine locations of telecommunications towers. *Hydrographic surveys* are used to map the bottoms of lakes, rivers, and other bodies of water, determining their physical shape and characteristics with respect to landscape. Engineers involved in the layout and measurement of roads or other human-made structures perform *construction* or *road surveys*. Often in forestry and agricultural applications, a survey is completed using aerial photography, where tone, texture, and shape are interpreted, delineated, and measured. Such surveys are referred to as *aerial surveys*. The use of satellite imagery is sometimes called a *remote sensing* or *satellite survey,* although the latter term is misleading because it suggests that the satellites themselves are being surveyed.

In many countries a census or *demographic survey* is used to determine the numbers and characteristics of a population with respect to a geographical area. The demographic survey provides information that is useful for determining distributions of populations. They are used for urban planning, transportation analysis, and determination of new store locations. The demographic is perhaps the most difficult survey to conduct and maintain. They are very expensive and by necessity must be kept up to date if they are to be useful; otherwise, they are only historical representations. A location-based application involving outdated demographics can lead to some very strange conclusions and results.

In 1784 in the United States, the Continental Committee initiated a means of locating and positioning legal land boundaries. This resulted in 1785 in the establishment of a coordinate system. The Dominion Land Survey of Canada (CLSS) has a similar coordinate system, consisting of townships and ranges. In Australia, the Geocentric Datum of Australia (GDA) is based on the Australian Fiducial Network (AFN), which fits into a global geodetic framework. The AFN comprises eight highly accurate survey marks across Australia, each with a permanently tracking GPS receiver. It has been established by the Australian National Mapping Agency for geodetic surveying and scientific purposes. The AFN was used as the foundation to determine geocentric coordinates for the Australian National Network (ANN). The ANN comprises 70 survey points across Australia spaced 500 km apart, each having latitude and longitude on the network. GPS data have been obtained at each of the 70 survey points, allowing ANN points to be linked into the framework provided by the AFN (AUSLIG, 2000). For Great Britain the British National Grid consists of a series of (100 × 100 km.) grid squares, each grid uniquely labeled using two letters. More recently, the British Ordnance Survey (OS) has adopted the *topographic identifier* (TOID), a unique 16-digit value that is related to landscape position and features.

A key difference between the systems of Canada and the United States is in the numbering of townships. In the United States the first township begins in the northeast, ending with the thirty-sixth township in the southeast. In Canada (CLSS) the first township begins in the southeast but ends in the northeast, running consecutively. In both countries a series of permanent benchmarks that have been accurately placed serve as monuments. These markers serve as survey control points or tie points. For a survey to be used with other survey information, both sets of data should logically tie to a common known control point or system.

Those areas that are surveyed with GPS today are then downloaded from a separate file. They may be integrated with already downloaded and stored spatial data files collected similarly. If the GPS information for the new and existing sets of data have similar projections and datum, the files can be merged. If one set differs from the other in either of these regards, the data should not be merged because the two sets to be combined do not have similar data integrity. A review of the metadata will include information about the datum as well as the projection used at the time of data collection.

While geodesy is focused on physical measurement of Earth's surface, delineation of boundaries is not necessarily as clear and easy in some surveys. This is due in part to Earth's shape. For small-scale surveying, that requires less accuracy; spheroid approximations are used to calculate Earth's shape. For large-scale mapping, ellipsoid approximations and calculations are more precise, due to the higher levels of accuracy needed.

Not all physical features on the landscape are always apparent. A hydrographic survey is a prime example. Some landscape streams may be intermittent, occurring irregularly during the course of time, flowing only when it rains and strongly influenced by antecedent soil moisture and slope. Therefore, whereas geodesy considers the measurement and determination of boundaries and is related to physical entities and positions, biological phenomena are not considered geodetic surveys, although they depend significantly on geodesy in their representation.

2.2 LONGITUDE AND LATITUDE

Coordinates are projected using two types of parameters. *Linear parameters* take into consideration measured distances and scales, whereas *angular parameters* are based on degrees and angles. The universal transverse mercator (UTM) coordinate system is an example of a system using linear parameterization, whereas latitude and longitude use an angular parameter. Spatial data may be transformed from one system to the other for application and use. *Latitude* is the angular distance as measured from the center of Earth to a point north or south of the equator. *Longitude* is the angular distance from the center of Earth as measured east or west of a point on Earth's surface. Usually, we consider that point to begin at a point in Greenwich, England. Longitudes extend around the Earth and arrive back at the same point in Greenwich. However, it is more common to extend points only halfway around the world, in other words, to measure points as 180° east or west of Greenwich, England. Those measured west are expressed as negative values (e.g., −113° longitude is Edmonton, Canada). Those points measured east of Greenwich would then become positive numbers (e.g., +13° longitude is Berlin, Germany). Lines of longitude that run perpendicular to the equator are called *meridians*. Lines of latitude that run parallel to the equator are called *parallels*. Meridians converge at the poles, and parallels never converge. Therefore, 1° of latitude and longitude represent different areal distances at different places on the Earth, due to the fact that angular measurement is involved and applied to an ellipsoid surface.

Not all GPS or GIS systems display longitude and latitude to the same number of significant digits. This can affect the accuracy of position locations when they are presented. The following are examples of latitude and longitude that various GPS and GIS devices display:

- 13°N 23°E (in degrees of latitude and longitude)
- 13°21′N 13°23′E (to minutes of latitude and longitude)
- 13°21′34″N 13°23′21″E (to seconds of latitude and longitude)
- 13°21′34.7″N 13°23′21.8″E (to tenths of seconds of latitude and longitude)
- 13°21′34.77″N 13°23′21.89″E (to hundredths of seconds of latitude and longitude)

Depending on where you are with respect to the geoid and whether or not a GPS navigator can record beyond minutes, including seconds and even hundredths of seconds, accuracy may be affected. This has very practical considerations for applications since a great amount of effort may be placed on locating a position within centimeters but the GPS navigator cannot record the location because it does not have the capability to record seconds. UTM positions may experience a similar problem in recording a position. The point to remember is that a GPS navigator is digital; it displays information and records it with respect to significant digits. Those that are able to display and record more significant digits will also be capable of performing more accurate positioning.

2.3 DISTANCE BY TIME

Latitude can be determined from the angle of the sun with reference to a position on Earth. Another option is to measure the distance traveled in relationship to time. Moving 3° per hour for 3 hours yields a 9° change in distance. Does this provide an accurate distance measurement? One could argue that it does over shorter distances at slower speeds. But over longer distances the shape of the Earth needs to be considered. The accuracy by which the time is measured would also be a consideration. What if you were on a jumbo jet traveling at 800 km per hour? Do you move the same distances as if you were to move perpendicular to the equator, or northwest by southeast? Longitude and latitude are measured as degrees of arc from the center of Earth rather than in time units such as seconds, minutes, hours, and days.

Keep in mind that angular measurement is denoted in seconds, minutes, and degrees (as is time). There are 65 Loran-C stations dispersed throughout the northern hemisphere. Chained together in groups of four or five, these stations provide accurate time since they are referenced to atomic clocks. Loran-C is particularly useful where coastal spatial applications are located. GPS is, of course, highly accurate with respect to time.

People sometimes collect GPS points in the field, download them, then try to orient them on existing maps. In some cases the GPS points do not align properly with the existing map. The reason for this is that the map may have been generated using measurements based on one datum as compared to the GPS collection of points, which used a different datum. All spatial data should necessarily attempt to take into account the shape of the Earth through the application of a datum. If all countries had their own time based on their own prime meridian, it would be difficult to compare one region to another with respect to time. The development of a standard time based on meridians takes into account that the Earth revolves 360° every 24 hours, which represents 15° per hour (360/24 = 15). Thus, one time zone would be equivalent to a distance of 15° of longitude. If the time zones were not referenced to a similar system of angular measurement, the determination of time over distance would require many changes and computations. Although possible, this would cause great problems in crossing North America east to west, where many adjustments would have to be made due to individual countries, states, and provinces having their own system of measurement.

In 1884 the International Meridian Conference met and decided on one longitude on which time would be based—Greenwich, England, which we now call the *prime meridian* (0° longitude). In 1928, Greenwich Standard (or Mean) Time (GMT) was replaced with a new universal time, which in 1964 was given the name *Universal Time Coordinated* (UTC). Those of you working with GPS know that GPS time is measured using UTC—it shows up on the GPS liquid-crystal display. When a GPS is turned on, the display will usually show the time using a 24-hour clock, or perhaps in both local and Greenwich time, depending on the type of GPS being used. The time zones are based on 7.5° on each side of the *controlling meridian* used to determine them and are considered to represent 1 hour in time. By now you have probably guessed that an airplane traveling at a constant rate will cover varying distances with respect to the equator for the same time, because 1° can represent different distances.

Let's go back now and look at representing real-world 3D positions in two dimensions, such as with a map. When locating positions on the Earth during fieldwork, positions ultimately will be translated into a 2D map. How does one take positions derived from an ellipsoid and present them accurately in 2D on a flat surface?

Consider, for example, taking an orange and laying its peel flat. Try peeling an orange carefully and then laying it on a piece of paper. No matter how the peel is removed, it is impossible to lay it down flat edge to edge over its entire shape without distortion. Try it!

Suppose that you have marked two positions on the orange when the peel was attached. Is the distance between them representative and accurate when the peel is laid flat? Probably not. This demonstration in its simplest form provides an example of the problem in trying to take an Earth position collected from a ellipsoid and representing it on a 2D surface in the form of a map. The process of making this kind of transfer is called *projection*—attempting to take a 3D object and represent it in 2D. Many projections have been developed and are in use around the world. Each projection has advantages and disadvantages with respect to accuracy when attempting to describe and portray positions from a round 3D world to a flat surface—a cartographic map. Looking again at the orange, it can be noted that most of the middle sections of the peel lay flat, whereas areas near the top or bottom (the poles) are shifted a fair amount.

Different projections are used for different applications and areas around the globe (ellipsoid), depending on the amount of tolerance permitted in the representation. For larger geographical areas (i.e., national) it is often desirable to use a cylindrical projection, which resembles the curvature of the Earth and is most accurate with respect to those positions on or nearer the equator (when the orange was laid out, you probably noticed that a large area in the middle remained intact). A projection imagined as a cone shape placed atop Earth's poles and radiating outward is called a *conical projection*. If the cone were to be unfolded, with the positions of Earth marked along the inside of the cone, the image would look like it radiates from the center or cone tip, similar to the arrangement in Figure 2.1. You will note that the parallels are evenly spaced and that the meridians are pointing to the north (the point of the cone) and are straight. Such an *azimuthal* or *planar projection* results in a map whose latitudes vary in width. *Azimuth* is the measurement of angles from the north or starting point of 0°, traversing a circle, and ending at 360° or 0°. *Bearings,* in contrast, divide a circle into quadrants. An azimuth of 185° is equivalent to a bearing of S 5°W. Using an azimuthal projection, any point on Earth can be determined from the north or south pole using azimuth and distance.

This method can also be used with other positions on Earth. But there is more. Perhaps you have had the opportunity to traverse with a compass. An *open traverse* begins at point A, moves across a landscape, and ends at point B—the traverse does not close upon itself. A *closed traverse* ends back at the originating point.

A GPS is able to provide both azimuths and bearings. If a series of waypoints are located, with the simple push of a few buttons the user can determine the azimuth or bearing from any given waypoint to another waypoint. Since both location and time are being recorded with the GPS, they form the basis for determining the estimated time of arrival—from waypoint to waypoint. Using a compass along with a GPS has other important practical applications, discussed in more detail in Chapter 5. Similarly, using an aerial photograph and determining north, both azimuths and bearings may be calculated since aerial photographs have a *nominal scale* (i.e., the average

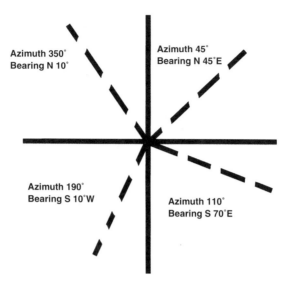

Figure 2.1   Azimuth and bearing.

scale). Thus, all locations on an aerial photograph are not of similar scale but vary with relief (i.e., scale is relative to the principal point).

Conversion from one coordinate system to another is required for many GPS/GIS applications, although many GPS users prefer UTM (Figure 2.2). Most GPS navigators provide UTM coordinates in meters (i.e., 12N–0256638–4015002). This string of numbers indicates that the location is in UTM zone 12 north and has an easting coordinate of 256,638 m and a northing coordinate of 4,015,002 m. This means that the given position is 243,372 m west of the central meridian (which has coordinate 500,000) for zone 12, and 4,015,002 m north of the equator (see Section 2.4). A look at a Mercator projection shows little curvature, with latitude and longitude displayed as straight lines.

There are numerous types of projections, the Mercator projection being one of the most commonly used. However, due to scale exaggeration, it can result in distortions and error. It is important when considering projections to understand that no map projection is without error. Each projection will have shortcomings with respect to distortion, which can be attributed to taking a 3D surface and portraying it in 2D. The following are general characteristics of four types of projections:

- *Azimuthal projections:* true at their center points and distorted increasingly toward the edges
- *Conical projections:* true along the parallels between the pole and equator but distorted increasingly at the poles and equatorial areas
- *Cylindrical projections:* true at the equator but distorted increasingly toward the poles
- *Universal transverse mercator* (UTM): design for optimal use from latitude 80° south to latitude 84° north (the poles are at infinity)

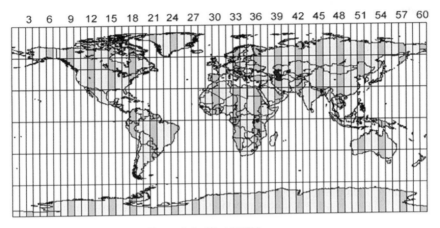

3 6 9 12 15 18 21 24 27 30 33 36 39 42 45 48 51 54 57 60

Figure 2.2 World UTM zones.

2.4 UNIVERSAL TRANSVERSE MERCATOR

Mapping requires that positions on Earth's surface be defined mathematically. There are several ways to do this including using latitude and longitude or *Cartesian coordinates*. Cartesian coordinates can include *x,y,* and *z* axes; longitude and latitude are angular measurements and each coordinate may also be assigned height, or a *z*-axis. Longitude and latitude are referred to as *geodetic* or *geographic coordinates*. For determining locations on Earth's surface, the ellipsoid shape of the Earth must be considered. There must be a reference point from which to begin, usually considered to be the center of the Earth's mass, and the ellipsoid dimensions are calculated from that point. When referring to a height or elevation, we are usually talking about a location above mean sea level, or on the ellipsoid. If heights were based above the ellipsoid, there would be differences in elevation worldwide for the same elevation, depending on the location, due to the relationship to the ellipsoid. Consequently, a datum is used to represent the ellipsoid more accurately for a particular area on the planet.

To aid the cartographer with the task of representing a 3D surface on a 2D surface or piece of paper, several projections have been developed throughout the world. One of these projections, developed by Gerhard Mercator in 1569, is the Mercator projection, known as a *conformal* or *orthomorphic projection*. Maps projected using a Mercator projection have straight lines for all meridians and parallels. Progressively larger distortions occur when moving toward the poles, due to the ellipsoid shape of the Earth. In 1772, Johann Heinrich Lambert invented the transverse Mercator projection using a spherical Earth shape. In 1822, Karl Friedrich Gauss used the ellipsoid shape, adapted in the 1920s by Leonhard Kruger (Langley, 1998). Using the transverse Mercator projection, the central meridian, equator, and meridians extending at 90° from the central meridian have straight lines.

Adjusting the central meridian slightly for scale results in two parallels on either side of the central meridian that are similar in scale. Thus, transverse Mercator zones are determined using a central meridian and a parallel on each side of that central meridian, and areas within those zones are reasonably close to each other in scale. UTM grid lines run perpendicular to each other and are measured in meters. The transverse Mercator system differs slightly from the Mercator projection in that in the transverse Mercator the cylinder is rotated such that it touches the 60° longitudes, which form the central meridians of the transverse projection. In the UTM system, distances are termed *northings* and *eastings,* both measured in meters. The northings begin at the equator and move north or south from the equator. Unlike meridians, parallels are unequally spaced in UTM. Eastings are distances from the centered meridians that run north–south.

In the UTM system there are 60 central meridians of 6° each that originate at a point of 180°. Working with GPS you will be familiar with UTM since most GPS navigators are capable of collecting data using UTM. Similarly, most GIS software used for mapping is also capable of working in UTM and/or transforming data from other coordinate systems to UTM or geographic coordinates. UTM northings are divided into 8° zones of height covering the entire distance from 80° south to 84° north latitude. The northing zones are labeled C-X, with M/N located at the equator. Positions around the world using UTM are all measured as positive numbers (in meters) where a system of false northing and false easting is utilized. Adding 500,000 m to all meridians and 10,000,000 m to positions south of the equator ensures positive numbers.

Map scale increases by a factor of 0.9996 to each side of the central meridian, which results in maintaining average map scale over longer distances from the central meridian. The meter was originally determined as 1/10,000,000 of the distance from the equator to the North Pole. Using the World Geodetic System (WGS84), the 45th parallel had a distance of 4,986,272 m instead of the mathematical one-half of 500,000 m. Using the North American Datum (NAD27), the northing value to the 45th parallel was measured as a distance of 4,986,055 m. This difference is attributable to the datum used, again providing an example of the importance of understanding Earth's shape.

An example of the universal transverse Mercator is the British National Grid. It consists of regularly spaced grids that measure 500 km by 500 km. These are then further subdivided into squares 100 km by 100 km and assigned coordinates. The squares are then assigned two letters and include coordinates for positions within the 100-km² grid. It should be remembered that any grid system may be used for local surveying and mapping. The grid system is then later tied in to a control reference grid system or geodetic network. Examples of this can be found in archaeological mapping. Using a local coordinate or grid system, very small grids may be established for the purpose of unearthing historical artifacts. Grids for this purpose may be below 1 m in size, and the entire area may not exceed 20, 50, or even 100 m altogether. Later, this small grid is integrated with the larger UTM grid that serves as a control for the site. Each of the smaller grids would be assigned UTM coordinates and perhaps even a local reference system much like the British National Grid. It could include whatever naming or identification convention archaeologists wanted to use.

Areal coordinates that are not tied into a geodetic network may be used in some applications. These coordinate systems often start at one point (0,0), are labeled locally, and may be in whatever units are desired. They are often used for small projects or research that does not depend on tying into national grids. All of the spatial information gathered for these projects is aligned to the same grid. Consequently, they can be transferred for use in a GIS, analyzed together, and represented.

2.5 DATUMS

A *datum* can be defined as a series of parameters and control points used to determine and define accurately the 3D shape of the Earth. The corresponding datum is the basis for a planar coordinate system. The objective is to project coordinates on datums. For example, the North American Datum 1983 (NAD83) is a common datum system used for maps and coordinates in North America. Around the world a larger number of people use the World Geodetic System 1984 (WGS84); this datum is common in most GPS equipment and is found in most GIS software. In the discussion of UTM it was noted that there is a difference of some 217 m when changing from WGS84 to a datum of NAD27. Datums are important because they provide a frame of reference for mapping to define a map's grid of coordinates. Selection of a particular datum can have a significant impact when navigating as well as the positional accuracy of a map. A cartographer will often first determine a projection to be used, which converts the spherical surface of the Earth to a flat map or 2D image prior to map construction.

Subsequently, mapmakers utilize models to approximate the ellipsoid shape. Datum models also reflect the local variations caused by the irregular surface of the Earth (Snyder, 1982). The geoid is the shape of an assumed Earth using a theoretical water surface (i.e., a surface completely at a theoretical sea level) as it would be without terrain and without external gravity (i.e., no spin, no tides). A geoid is therefore a smooth surface close to real mean sea level (MSL $\pm$ 40 m) over the entire Earth, variations depending on the gravity effects of mountains, trenches, and density. The exact geoid shape therefore varies with locality—again because the Earth is not round. For accurate local mapping, it is necessary to match the local geoid to a formula that most closely approximates the Earth's shape at the location being considered. Heights above or below a mean sea level reference are termed *orthometric heights.*

Users of geospatial data also need to know the coordinate system and projection that the data are in so that they can make accurate measurements from the data or overlay different data sets for later analysis. In the United States, NOAA has been moving from use of the North American Datum 1927 (NAD27) to the newer World Geodetic System 1984 (WGS84) standard. In Europe many different datums are in use, depending on the country. In other parts of the world, charts are based on other formats, such as WGS72 or local or regional datums. Two different datums may range from 40 m apart in Florida to 150 m apart in Maine. This is one reason that metadata are important; they provide a record of the datum and projections used for a given data set. Thus, metadata provide a means for integrating one set of data with another, thus ensuring conformity and accuracy.

A GPS stores datums from all over the world, making it possible for the user to select the correct datum for the region under study. GPS users should become knowledgeable about the datum in their area and check their GPS navigator to ensure that the relevant datum is present. Failing that, data may still be recorded, but the datum used should be noted so that the user can later convert to the proper datum and then transform the data using mathematical algorithms designed for such transformation.

To understand the relationship of datum more fully, GPS users can perform a short exercise. Stand in one location, record a waypoint, then record the same waypoint labeling it differently using a different datum. Do this for six or eight different datums. Download the information and transfer it to a GIS and look at the various waypoints that have been produced using differing datums. Assuming that the information is differentially correct, the positions will probably not line up; instead, they will probably be some distance from each other. Those differences are due largely to datum selection, particularly if they exceed 37 m, since that is the minimum error for GPS 95 percent of the time. For a more interesting comparison, try this same exercise at higher and lower elevations. Since datums are designed with a view to mean sea level, any change in elevation above or below this level can also introduce errors associated with the datum application. That is, a datum is calculated based on an average shape for the Earth in a local region. It is important to remember that position errors may have high–low accuracy, high–low precision, or perhaps be biased.

2.6   SCALE REPRESENTATION

Whether we are measuring a distance on the ground manually with a tape, using a GPS navigator to acquire positions between two points, or using aerial photography to measure the distance between two points, we are interested in scale, more accurately, *map scale*. Map scale can be defined as the measure on a map and its equivalent measure on the Earth, often expressed as a representative fraction of distance, such as 1:70,000. This means that 1 unit on the map represents 70,000 of the same unit of distance on the Earth. There are various methods for expressing this relationship, including:

- *Verbally:* One centimeter on the map equals 100 km on the landscape.
- *By representative fraction:* 1:100 (note that units are not used)
- *Graphically:*

  0   100   200   300   400   500
  |__|__|__|__|__|__|__|__|__|__|__|

In the British Commonwealth a scale commonly used is 1:63,360 (1 map inch equals 63,360 of the same unit). The reason this scale is commonly used is because there are 63,360 inches in a mile. This scale has a benefit in terms of ease of understanding because 1 inch on a map is equivalent to 1 mile, ½ inch is equal to 0.5 mile, and so on. This scale is also easily transformed to various other scales which are multiples of 63,360, including 1:125,000, 1:250,000, and 1:500,000. In the United States the U.S. Geological Survey (USGS) now employs a scale of 1:24,000, for the reason that 1 inch can be used to represent the equivalent of 2000 feet in actual ground distance. Countries that use the metric system often use scales of 1:10,000, 1:20,000, 1:50,000, and so on.

In fact, a cartographer can produce a map at any scale. Selection of scale depends on the detail of information to be represented. *Large scale* refers to maps whose scales are 1:50,000 or greater (1:30,000, 1:20,000, 1:10,000) or any scale moving toward 1:1. Notice that greater scale means smaller representative fractions. On large-scale maps the user can see increasingly more detail on the map. *Small scale* refers to maps whose scales are less than 1:50,000 (1:60,000, 1:80,000, 1:250,000, or any scale to infinity). On small-scale maps the user will see less detail but a far larger area than on a large-scale image or map. Scale is also important when using aerial photography. Aerial photographic stereo pairs have varying scales depending on the flying height and focal length of the camera used and the terrain relief—flat land will have a more similar scale.

Sometimes large scale and small scale are confused. The easiest way to distinguish one from the other is to think in terms of the objects being viewed. Large (close) objects indicate large scale, small (distant) objects indicate small scale. If a map of an urban subdivision is required, perhaps a large-scale map (1:1000 or even greater) might be necessary to see the details of the neighborhood. Alternatively, forest managers and those interested in large-area landscape projects would be interested in mapping very large tracts of landscape requiring small-scale maps (1:20,000 or smaller).

The U.S. Geological Survey and most countries publish maps at various scales. The scale used for most U.S. topographic mapping is 1:24,000. Maps published at this scale cover 7.5 minutes of latitude and 7.5 minutes of longitude and are referred to as *7.5-minute quadrangle maps*. Map coverage for the United States has been completed at this scale, except for Puerto Rico, which is mapped at 1:20,000 and 1:30,000 and a few states that have been mapped at 1:25,000. A 1:24,000 scale is fairly large. A map at this scale provides detailed information about the natural and human-made features of an area, including the locations of important buildings and most campgrounds, caves, ski lifts, watermills, and even drive-in theaters. Footbridges, drawbridges, fence lines, private roads, and changes in the number of lanes in a road may not be shown, though.

The level of detail decreases moving from large scale to small scale. Features become less apparent, and this is termed *generalization* (see Chapter 4). Generalization results in information being eliminated from maps in the 1:50,000 to 1:100,000 or smaller scale range. These maps cover more area while retaining a reasonable level of detail. Maps at these scales most often use the 15-minute or 30- by 60-minute quadrangle formats. The smaller the scale becomes, the more generalized the view of the landscape; that is, the fewer details the user will be able to see. Those who have worked with a GIS will have seen this effect, where scale changes in maps seemingly change the level of detail viewable as new maps are designed and displayed, particularly when maps are zoomed in and out.

Such effects are also attributable in part to hardware factors. Graphics cards render different numbers of pixels, and a digital map is composed of pixels. If the graphics card is unable to provide a resolution of 1600 × 1600 dots per inch, no matter how close you might want to zoom in, you cannot exceed the capabilities of the graphics card (more detail) and remain at the same level of detail—a factor often overlooked. For the same reasons, computer game software usually calls for an upgraded graphics card to render quick response in numerous colors and at higher resolutions.

It is also important to understand from a GIS perspective that data collected in one scale may include details for one theme which are not present in other thematic layers (variable) collected at the same or another scale. There are two reasons for this. First, data collected at a similar scale for the second thematic theme may be of a different type, which is why they are being integrated thematically, so that a more complete map and representation of spatial information for the area can be produced. An example of this is the collection of data for a large lake and a small road. The lake would be readily apparent at a scale of 1:30,000, whereas the road might be nearly invisible at the same representative scale.

In the case of one thematic layer collected at one scale being integrated with another layer (variable) at a different scale, the theme with the lower scale (i.e., high representative fraction, or small scale) will have a higher level of generalization. Thus, when lowered in scale to the same scale as the first thematic layer, a line, for example, looks very wide and perhaps very straight rather than more closely following the terrain shape and landscape forms of the first thematic layer. The mixing of data collected from two different scales can present difficulties when attempting to view higher levels of detail. The solution to this problem is to re-collect information at a scale that is more similar to the scale of other thematic layers.

Since GIS and other technologies are in their infancy, data conversion and integration concerns surrounding transformation remain common. Many organizations are still upgrading from legacy systems. This involves the mixing of data from numerous sources at different scales along with associated difficulties arising from varying formats. This again supports the need for metadata that would provide information about the characteristics of a given data set. At the same time, client architectures are becoming more readily available that are capable of accessing data from remote servers regardless of data format. This is achieved by clients connecting to remote servers comparing metadata, allowing for a higher level of interoperability. Although this may permit the integration of data sets for similar regions, a concern arises as to the scale of each layer being integrated. It is important that the user of spatial information understand that information collected at one scale is usually best represented or integrated with information collected at or near a similar scale. As an example, take a USGS GTOPO digital elevation model and try to apply it at 1:10,000. Since GTOPO is based on a very small scale grid, it is not possible to detect minor changes in topography in local regions. It does, however, permit the representation of topographic change over larger regions and continents.

2.7   ACCURACY AND PRECISION

Accuracy is a measure of how close a point is to its true position (Figure 2.3). *Precision* refers to how closely measurements are in relation to each other. Both accuracy and precision may be affected by *bias,* a systematic error that affects all data being collected or represented. As an example, a landscape position and associated features

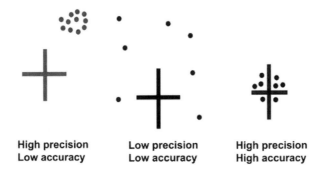

High precision   Low precision   High precision
Low accuracy   Low accuracy   High accuracy

Figure 2.3 Accuracy and precision.

were located with a GPS navigator. A map was produced from the GPS data and the locations were later "ground-truthed" and compared to the known legal survey coordinates and found to be accurate, that is, to be where they are truly located. In this case the GPS survey would be deemed accurate. Assume that the GPS survey was conducted a number of times and that when compared to the legal survey, the positions were found to be within 1 m or so (closer is better). These points are all accurate and have high precision since the positions are clustered together and surround the true known legal survey coordinates. Using the same example, let's assume that a single GPS waypoint was taken and compared to the legal survey coordinates. This position was found to be a distance of 10 m away from the known legal position— less accurate. If 20 readings were then taken using GPS, all found to be near each other some 10 m from the known legal position, the 20 readings have high precision but lower accuracy. Finally, let's assume that a GPS survey is made and the positions are all clustered closely together but at some distance from the known survey point. Furthermore, all our GPS surveys show the same consistency, each location about the same distance from the known legal survey point. This would be bias. Bias can appear in GPS information when, for example, the GPS is using the wrong datum for the part of the world where the waypoints are being taken. A datum is optimized for particular locations around the world assuming the shape of the Earth within the local region.

The scale of a map or photograph leads to other important practical considerations relating to accuracy. In most cases a pen or pencil tip is about 1 mm in diameter. On an aerial photograph at a scale 1:10,000, could you pinpoint a GPS location accurately on the aerial photo? Keep in mind that a GPS has an accuracy of about 1 m or better when corrected differentially. Would you increase your chances of locating the point on the aerial photo if it were a larger- or a smaller-scale photograph? Are there other considerations? Using this example, 1 mm is equivalent to 10 m as measured on the ground, meaning that the smallest identifiable unit that can be seen accurately is 10 m in size. The GPS is, however, able to locate a point on the image to within 1 m. It would not be possible, for example, to apply the GPS location accurately to a 10 × 10 m swimming pool. Which 1-m portion of the 100-m^2 swimming pool would receive the GPS coordinate? This exemplifies the issue of GPS—image accuracy—

leading to important questions when georeferencing a digital image, which we discuss in more detail in Chapter 8.

There are other factors that need to be addressed when GIS is considered in this example. Assuming that the image is to be used as a *backdrop* in a GIS, it must somehow first be georeferenced. A backdrop is a map or usually a photograph image placed as the bottom-most thematic layer in a GIS, serving as a reference for other thematic layers placed above it. The reason for having a backdrop is to provide a photo-realistic representation theme, allowing locations to be easily seen and identified with other thematic layers. Backdrops may also be used for digitizing where on-screen digitizing takes place. *Georeferencing* means that the image will have a coordinate system of some kind associated with the image, enabling areal distances and angles to be determined accurately. Once the image or backdrop map is georeferenced, a scale will be available. The scale will allow distance between any two points on the image to be determined while using the GIS software.

If themes are to be compared and analyzed, they must be georeferenced similarly. Consequently, the datum used while collecting GPS point data has direct implications with respect to later GIS analysis and accurate distance measurements involving other thematic information. It is important that data collected with a GPS navigator using a specific datum use a similar datum when rendering the data image. One must be careful not to make the mistake of taking GPS data collected in one datum, displaying the information in another, then merging that information to a third datum.

Transformation or *conversion* refers to taking one set of geospatial data, collected and georeferenced using one datum, and changing it to another datum (e.g., NAD27 to WGS84). The GPS information and datum should match the GIS datum being used before analysis takes place. Integrated geotechnology applications require close attention to, and the development of, metadata. It is not uncommon to be using three, four, or more geotechnologies that utilize different coordinate and datum systems. Keeping track of their respective datum is critical for ensuring accuracy and decreasing error propagation throughout an integrated data set. Most data conversion of spatial information is either from raster to vector format or from vector to raster format. This usually involves either digitization or scanning. In a GIS these raster formats are sometimes called *grids.*

Interesting work continues internationally in an attempt to develop methods for measuring accuracy and performance error modeling within GIS. Uncertainty modeling has been used to assess position quality, and metamaps are presented visually which provide an indication of the level of topological errors (Hunter, 1999). Map accuracy also involves the perception of mapping products. As one author points out (Monmonier, 1996), there are many different ways to lie with maps. What maps represent through the use of color and other cartographic elements affects their attractiveness and communication usefulness. Perhaps the largest issue surrounding map accuracy has to do with timeliness. Analysis performed using old maps and data, particularly for urban applications, is subject to landscape changes due to urban development and growth. Similarly, location accuracy may change over time as entities themselves change: for example, rivers, streets, and building size and shape. The use of up-to-date spatial information, particularly for biologically related modeling, requires up-to-date information.

2.8 COORDINATES

The Dominion Land Survey System originated in western Canada in 1870. Now called the Canada Land Survey System (CLSS), this system was based on meridians, baselines, townships, ranges, sections, and subdivisions of sections (Figure 2.4— McKercher & Wolfe, 1986). The system begins at a point some 15 miles west of Winnipeg, Manitoba, Canada (97° 27′28.41″). Principal meridians extend every 4° in a westerly direction across western Canada, extending to the province of British Columbia and finally to the Pacific Ocean. In British Columbia a special coast meridian exists (122° 45′36.6″). From the principal meridians a series of townships and ranges that form a grid coordinate system are derived. There are 36-square-mile townships, each divided into quarters of 160 acres and further divided into legal subdivisions.

Interestingly, in the CLSS a series of *correction lines* occurs periodically. If you live in a rural Canadian community, you will come across these correction lines when driving from place to place. These lines are necessary because the system is based on longitude and latitude, which relate to meridians and ranges. Remember that latitude and longitude are spherical coordinates, based on angular measurements. As mentioned previously, when moving north the lines of longitude converge, which accounts for the coordinate grid not remaining square in the CLSS, due to the shape of the Earth. Therefore, correction lines are used to readjust the coordinate system every second township when traveling north or south. This results in a jog, with north–

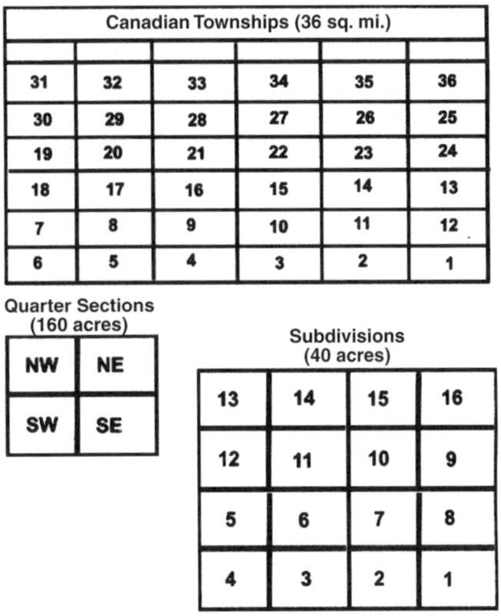

Canadian Townships (36 sq. mi.)					
31	32	33	34	35	36
30	29	28	27	26	25
19	20	21	22	23	24
18	17	16	15	14	13
7	8	9	10	11	12
6	5	4	3	2	1

Quarter Sections
(160 acres)

NW	NE
SW	SE

Subdivisions
(40 acres)

13	14	15	16
12	11	10	9
5	6	7	8
4	3	2	1

Figure 2.4 Township coordinates: Canada Land Survey System.

Table 2.1 **(x,y) coordinate system**

Object	x-Coordinate	y-Coordinate
1	3	4
2	6	5
3	8	7
4	2	2
5	5	3
6	1	1
7	9	9
8	7	6
9	3	8
10	2	7

south roads offset. The Canadian Land Survey System is an example of a *coordinate system*, a reference system on which spatial information can be referenced to determine location. The simplest coordinate system remains the (x,y) planar coordinate system, which is useful for locating any point (see Table 2.1). This coordinate system is easily constructed using a spreadsheet, or it may be a series of GPSs downloaded from a GPS navigator (see Table 2.2).

Since each location or waypoint has an (x,y) coordinate, the waypoints can be plotted on a grid (Figure 2.5). In this case, the grid that is used is (10×10 units). This

Table 2.2 **(x,y,z) GPS coordinates**

Location	x-Coordinate	y-Coordinate	z-Coordinate
1	8435.406	876546.000	2345
2	8487.719	876549.500	2365
3	8276.875	876477.000	2387
4	8332.438	876484.500	2398
5	8384.000	876488.000	2376
6	8439.063	876491.000	2300
7	8487.813	876492.500	2310
8	8276.531	876427.500	2320
9	8332.188	876430.500	2341
10	8385.844	876733.500	2355
11	8485.406	876946.000	2359
12	8687.719	876349.500	2400
13	8246.875	876677.000	2430
14	8532.438	876784.500	2444
15	8304.000	876888.000	2378
16	8739.063	876891.000	2394
17	8687.813	876792.500	2397
18	8666.531	876627.500	2356
19	8332.188	876830.500	2345
20	8685.844	876733.500	2333

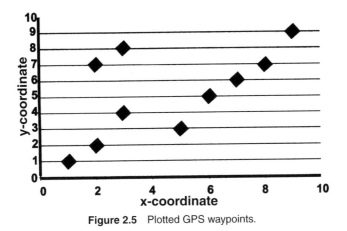

Figure 2.5 Plotted GPS waypoints.

plot permits a quick evaluation of the plotted GPS relative to their coordinates. The data could be exported in database format for use in a GIS. Using a GIS, these points could be analyzed, for example, for *proximity*. Proximity is a measure of distance between the individual waypoints—whether or not they overlap. Alternatively, areal distances between individual waypoints could be plotted, in meters in the case (shown in Figure 2.6). A coordinate system (x,y,z) would include a z-coordinate representing information in a third plane relative to the (x,y) coordinate. With the third coordinate the data set moves from being a 2D to being a 3D representation.

Figure 2.6 Proximity of GPS waypoints in GIS.

If visualization is the purpose of collecting GPS data, consideration regarding how the visualization will be represented should be considered. Data represented in a planar projection may not be clear in a 3D visualization environment. Virtual Markup Language (VRML) is a method for representing spatial information in 3D. Using VRML, objects may be zoomed in or out, tilted, or turned. From alternate angles it is sometimes difficult to see 3D relationships, particularly with few waypoints. Therefore, recording more GPS waypoints will often result in a more easily understood 3D rendering in VRML.

Representing waypoints of a homestead on rugged terrain from 2D GPS waypoints does not provide a means to perceive the terrain associated with the waypoints. Instead, a 2D planar map is produced that rotates when placed into VRML. Is that useful? The individual data points can be seen from various angles in the VRML representation, but in reality, given rugged terrain, one point may not in fact be directly observable from another; it may be obscured by hills or by vegetation or other objects. In fact, when a 2D contour map is transferred to VRML, it becomes more confusing to interpret the representation. Simply changing the perspective of the contours may even cause one to believe that distances and elevations as presented are larger or smaller than they really are, due to the orientation and viewpoint. Some spatial information is viewed more effectively in a 2D flat environment, whereas other data are more suitable for 3D representation. Consideration of the nature and characteristics of what is to be rendered is warranted (see Chapter 9).

A coordinate system does not have to be horizontal, parallel to Earth's surface. In some applications, such as monitoring noise levels along transportation corridors, vertical coordinates are useful. With GPS coupled to instrumentation, noise levels can be monitored using specially designed sound meters which measure the incidence of noise and the level in decibels, while the GPS would reference those attributes to coordinates. For example, noise levels can be measured from a raised road or tram moving through a residential area. The distance and height of surrounding buildings away from the noise source will decrease as the noise-monitoring equipment is raised in height from ground level. The intensity of sounds may be low to moderate in terms of decibels, thus not traveling or radiating far, but to plot them using horizontal GIS would indicate that all space surrounding a source area is equally noisy. In three planes (x,y,z) that may or may not be true, since sound is highly directional and will be affected by the acoustics of nearby surfaces.

For such applications the development of vertical thematic layers may be more useful. This is fairly straightforward, requiring that the coordinates be transposed, where either the (x,y) coordinate in turn becomes the z-coordinate. The probability that an object in a selected unit is observed—whether seen, heard, caught, or detected by some other means—is termed *detectability* (Thompson, 1992). Viewed in 3D, the levels of detectability are more readily identified. Geoscience professionals by definition are not only interested in representing spatial information but also consider scale, character, and processes of those pieces of spatial information they seek to analyze and present if they are to represent them properly. Just as some sensors and technologies are capable of acquiring very fine resolution or very coarse results, their effectiveness for use depends on understanding the phenomena they attempt to measure.

This is true for monitoring sound and also applies to the representation of many types of biological phenomena.

Another example involving coordinates is related to perceptions. An employee works for a courier service company and is responsible for delivering mail by bicycle in downtown New York. A GPS is attached to each bicycle for the purpose of identifying the locations of delivery people and the customers. This information is useful for targeting advertising campaigns and promoting efficient service. As the bicycles move throughout the city, their positions are located and later downloaded and plotted. One employee's bicycle has traveled 247 km this week, a fact about which she is quite proud. Almost every other employee traveled twice that distance. Management discusses the difference with the employee, but she is at a loss as to why the difference is so large. An examination of records reveals that this employee delivered three times as many letters to various customers as did the other employees.

Management is confused, since the distance this employee traveled is so much less. As it turns out, the employee parks her bike often, preferring to run between close buildings, reasoning that it is quicker to do so than to continually lock and unlock the bike and wait for traffic. In this case the GPS has correctly determined the positions of the bicycles and assigned them to grids. Using the grids, neighborhood statistics are determined which are based on the stops and coordinate; nothing relates directly to to the locations of individual customers. In fact, the number of customers served by the employee is significantly larger than the individual bike locations would suggest. The employee actually served six customers on average per stop, whereas the other employees served three, so it is no wonder that she traveled half the distance of her fellow couriers.

Taxicabs installed with a GPS provide another example. Locating where a taxi is assists in routing and assigning it to customers quickly. This is a very practical application that can increase quality of service, decrease fuel use, and apportion vehicles to specific areas of a city. The GPS provides information about where cabs are traveling, and a central dispatch logs the calls to individual cabs. From those pieces of information a number of questions can be addressed: identifying the area with the highest number of calls, where most of the cabs are located, who has been assigned calls, how many are assigned, how far the average fare travels, and the time when the calls arise. During rush hours or traffic delays, cabs can be rerouted or assigned alternate paths to avoid delays. Knowing the location of taxicabs is therefore very valuable information.

One piece of information that is not included, though, is how many people are in each cab per trip. That information would serve two purposes. First, it can provide an indication of numbers traveled per kilometer by area. Why is that important? If you were an advertiser and knew that many cars had many people in them, on average traveling for a few kilometers, you might decide to target cabs with advertising. Essentially, there are several groups of four people in a closed environment for a short period of time. Advertising campaigns could be tailored and based on activities located within nearby coordinates. Knowing that there are 3.6 minutes on average of four people's time for numerous cabs per day is valuable data. Second, these numbers are useful due to the fact that future business is influenced by word of mouth.

Higher numbers of riders translate into a higher potential for advertising. This example could be used for transit and highway systems. Knowledge of independent variables and their numbers can provide unique opportunities.

2.9 DATA FORMATS

Geospatial information is available in numerous formats, many of which have evolved from proprietary products and are therefore related to both hardware and software. In addition, mapping organizations at local, national, and international levels have developed formats for spatial data exchange. Most users of spatial information in North America, for example, are interested in acquiring spatial information that is provided by the U.S. Geological Survey (USGS). The USGS is responsible for assembling and making available various national map products. These include:

- *Digital raster graphics* (DRGs): scanned images of a USGS standard series topographic map, including all map color information. The image inside the map neatline is georeferenced to the surface of the Earth and fitted to the universal transverse Mercator projection. The horizontal positional accuracy and datum of a DRG matches the accuracy and datum of the source map. The map is scanned at a minimum resolution of 250 dots per inch.
- *Digital line graphics* (DLGs): vector representations from existing maps and other spatial data. They come in three types: large scale (1 : 20,000), intermediate scale (1 : 100,000), and small scale (1 : 2,000,000).
- *National Elevation Dataset* (NED): seamless 30-m digital raster elevation data covering the conterminous United States, Alaska, Hawaii, Puerto Rico, and Virgin Islands.
- *Digital elevation models* (DEMs): raster grids of regularly spaced elevation values derived primarily from the USGS topographic map series.
- *Global 30 Arc-Second Elevation Dataset* (GTOPO30): global 1-km digital raster data derived from a variety of sources.
- *HYDRO1K:* global hydrologic database derived from 1996 GTOPO30 data.
- *Digital orthophoto quadrangle* (DOQ): computer-generated image of an aerial photograph in which image displacement caused by terrain relief and camera tilts has been removed. It combines the image characteristics of a photograph with the geometric qualities of a map. The standard DOQs produced by the USGS are either grayscale or color-infrared (CIR) images with a 1-m ground resolution; they cover an area measuring 3.75 minutes in longitude by 3.75 minutes in latitude, approximately 5 miles on each side. Each DOQ has between 50 and 300 m of over-edge image beyond the latitude and longitude corner crosses embedded in the image. This over-edge image facilitates tonal matching and mosaicking of adjacent images. All DOQs are referenced to the North American Datum of 1983 (NAD83) and cast on the universal transverse Mercator (UTM) projection. Primary (NAD83) and secondary (NAD27) datum coordinates

for the upper left pixel are included in the header to allow users to spatially reference other digital data with the DOQ (USGS, 2002).

In Canada, the national government maintains small-scale mapping for the entire country and the individual provinces maintain intermediate- and large-scale spatial data.

- *Canada Land Inventory* (CLI): available in large-scale, $< 1:50,000$; intermediate scale, $1:50,000$ to $1:250,000$; and small scale, $1:1,000,000$.
- *Landsat 7-1G:* a radiometrically and systematically corrected LOR image. The correction algorithms model the spacecraft and sensor using data generated by onboard computers during imaging events. The radiometrically corrected pixels are resampled for geometric correction and registration to an Earth location with a geodetic accuracy 5 to 25 times the sensor ground instantaneous field of view (Natural Resources Canada, 2002a).
- *Atlas of Canada base maps:* provide coverage of all of Canada. Data elements are feature coded and structurally clean. Base map components are available in five scales and a number of data exchange formats. The $1:2$ million and $1:7.5$ million scales are the primary bases for all Atlas products. The $1:3,000,000$ scale data have been generalized from the two larger scales (Natural Resources Canada, 2002a).
- *National topographic map:* $1:50,000$ scale topographic map ideal for recreational activities such as cycling, canoeing, snowmobiling, fishing, camping, and hiking. Included are hills, valleys, lakes, rivers, streams, rapids, portages, trails, and wooded areas; major, secondary, and side roads; and all human-made features, such as buildings, power lines, dams, and cut lines. A $1:50,000$ scale map covers an area of approximately 1000 km^2 (Natural Resources Canada, 2002b).

As in the United States, several other sources of spatial data products available in Canada both privately and publicly. In both countries the states and provinces maintain their own sources of spatial information that is often available at larger scales.

In Great Britain, the Ordnance Survey (OS) is responsible for providing the national mapping services for the country. Unique to the Ordnance Survey is the *OS Master Map,* really a system of maps organized by topic, although individual topic maps can be produced. A user queries the master map database, which consists of several layers of mapping products, such as buildings, roads, tracks and paths, rail, structures, land, water, administrative boundaries, terrain and height, and heritage and antiquities. The data are then integrated and can be provided in digital format. This is a scalable system capable of providing both large- and small-scale products, depending on user requirements.

The British National GPS Network is also tied to the Ordnance Survey. This GPS network is capable of providing three different coordinate systems. The OSGB36 is the National Coordinate System used by the Ordinance Survey for all mapping products. Ordinance Survey Newlyn is used to provide reference for height measurements above mean sea level. In Europe, the most common coordinate system in use is ETRS89. The Ordnance Survey works with all three coordinate systems and each can be converted and linked back to the OS Master Map. In practice, this means that GPS users can perform GPS surveying and tie transparently to master maps in real time wherever they are in the U.K. This level of integration provides a means to geocode data directly as they are captured and to update existing databases. As a result, the OS Master Map is being updated continually daily with several thousand new locations and attributes.

2.10   THEMATIC INCONGRUENCE

The integration of geotechnologies involves the capture of information through the use of different technologies. Accordingly, technologies are selected for various purposes. Each may generate information with different scales and resolutions. Some phenomena can only be monitored or measured using a specific technology. For example, to study large tracts of land, satellite images may be preferred due to cost, timeliness, and consistency. Alternatively, smaller land areas might be assessed using aerial photography and photo interpretation because interpretative clues provide a rich source of otherwise unobtainable information. A key factor in deciding on which technology to use involves understanding the purpose and nature of variables and processes that are being observed. Cost is also a consideration, as is timeliness and quality.

Often, more than one technology may be used. Two, three, or more geotechnologies may be integrated for use in one project. Examples of this can be found in archaeological applications. Historical applications invariably involve using historical maps that must be digitized, GPS for identifying on-the-ground benchmarks, coupled to aerial photography or remotely sensed images.

Thematic incongruence is not a measure of registration consistency between thematic layers. Nor is it focused only on scale, since very inaccurate data can exist even at 1-m resolution. As mentioned previously, bias and precision must be considered. Instead, it is an understanding that involves scale, cost, application, usefulness, and quality differences arising from the merging of spatial information using integrated methods and geotechnologies. The key word here is *usefulness,* which implies quality and therefore accuracy. It also implies that the data are applicable to the application at hand. Low congruence would imply that even if the data have 1-m accuracy, they may not be as useful: for example, collected at the wrong time using incorrect sensor technology. One could collect hydrological information once a day with a highly accurate GPS but because of the low sampling interval, that information is less useful for understanding the hydrological characteristics present. As an example, intermittent streams may run only during those periods of high rainfall. Since most high-intensity rain showers are of very short duration, these areas will require monitoring when it is raining. High congruence implies that integrated data can be used for the application at hand.

Thematic incongruence represents incongruity of data, considering, scale, application, process being observed, representation, cost, and usefulness for the use and application of geotechnologies for integrated spatial studies where timing is also considered. Assume that a project is using GPS data with an accuracy of 1 m, satellite images that are 30 m in resolution, and aerial photographs at a scale of 1 : 10,000. The combined data set would be incongruent, each value having a different scale. Moreover, the data consist of both vector and raster information. A lower aerial photograph scale would result in more congruence. Thematic incongruence is directly related to the application; therefore, there are no charts or tables to act as a guide. The values are comparative measures between data sets for similar applications.

EXERCISES

2.1. Is 1° of longitude the same distance for all places on the Earth? Explain your answer.

2.2. Compare 1° of longitude for Miami, Florida with 1° of latitude for Sydney, Australia. Discuss any differences.

2.3. If GPS waypoints are collected using an even grid spacing of 100 m, what level of generalization might you expect? Discuss.

2.4. What is large scale as compared to small scale?

2.5. Name three methods of expressing scale.

2.6. Compare the UTM coordinate system with latitude/longitude coordinate systems, and indicate advantages and disadvantages.

2.7. Your task is to measure the phosphorus level in a small lake for a week. Using a GPS, how would you approach this task, the goal being to represent the phosphorus changes during the week?

2.8. Mean sea level varies due to tide and gravity effects. Using a GPS, do locations closer to oceans have higher levels of accuracy than those of locations farther from oceans? Explain.

2.9. Name the purpose and discuss the usefulness of a geodetic network.

2.10. You own a vineyard with the objective of producing very fine wines. What approach to meeting that goal would you choose using geotechnologies? Identify some of the spatial considerations.

2.11. A forest animal has a GPS collar that samples the animal's location once a day. Discuss how you would represent this information with respect to scale, and why you chose the scale you used.

2.12. Explain *declination.*

2.13. Discuss *thematic incongruence.* Provide an example in your discussion, outlining major points.

3

CARTOGRAPHY, MAPPING, AND MAP SERVING

3.1 INTRODUCTION

Map products may or may not include data collected using modern geotechnologies. This is particularly true for most mapping prior to 1990 or so. Although aerial photogrammetry has been used for well over 100 years, digital photogrammetry is relatively new. From a hardcopy pocket map to a large wall map, to a hand or computer-generated drawing, to maps provided by map servers that are telecommunicated around the world, maps generated from digital or nondigital technologies each include elements of cartography. The map is a representation of reality, in color or black and white, and 2D, 3D, or 4D. A map might represent a backyard, neighborhood, city, country, or the world, or might constitute a pattern of microbes or blood vessels within the human body.

Maps are used primarily for navigation and locating positions. They are also useful for communication, providing visual cues with respect to surfaces (3D/4D), indicating patterns of population growth, transportation corridors, business demographics, or urban sprawl. Some maps are social in nature; others may be used for military purposes. Others may exist for a short time, such as emergency maps, or involve long time periods (centuries or longer) and are created to present historical events.

Maps are one of the few things that most people probably use at least once in their lifetime. Whether buying a home, which requires a land registry title, or driving through an unknown location and getting lost, people turn to maps for guidance, direction, understanding, and knowledge. We learn much about ourselves from maps: How much income we have and where it is distributed, how many people live in a particular area, the incidence of crime, the location of rivers and lakes, where schools are, and environmental factors and their impacts are among many things that can be represented with maps. Children are educated with maps, as are adults. The blind and the physically handicapped use maps. Some maps are confusing, others are humorous; most are useful, although a few are not, and some are plainly incorrect.

A map may be in digital or hardcopy form. It may be large and fill several square meters or small and be placed in a wallet. There is evidence that the art of mapmaking evolved independently around the world and that the Chinese may have been producing maps as early as the seventh century B.C. The oldest example of a map dates back to around 6200 B.C. Although it may not meet the standards of modern cartography, a clay tablet provides evidence of a preliterate form of communication and describes a location on Earth. In the ancient ruins of Ephesus in Turkey can be found maps engraved in stone, believed to be the first instances of mapping and advertisement.

The art of mapmaking evolved through the centuries, increasing in detail and accuracy. Different techniques for measuring and recording locations on Earth aided in this universal development. The Greeks understood that the Earth was a sphere and were able to calculate the circumference. Ptolemy proposed a set of projections and coordinate systems that are still used today. Mercator created a map that allowed sailors to cross the oceans, and Newton's theories postulated that the Earth is not a true sphere but an ellipsoid. All these scientific achievements in the spatial sciences contributed to the evolution of cartographic advancement around the world.

Today we have aerial photography, electronic distance-measuring instruments, global positioning systems, remote sensing, and computer technology to measure the Earth and to provide cartographers with very accurate information to use in creating maps to meet many needs. "The character and technology of mapmaking may have changed over the centuries...but the potential of maps has not. Maps embody a perspective of that which is known and a perception of that which may be worth knowing" (Wilford, 2002). As Wilford states, although map technology may have changed over the centuries, the potential of maps has not, and the same can be said of the skill and expertise of mapmaking. These abilities are still needed, whether a map is hand-drawn or computer generated. Modern cartography may incorporate and utilize the most efficient high technology available, but the skill of the cartographer is needed to bring out the potential of any map.

There are many fine examples of historical maps available dating back to the early fifteenth century (Shirley, 1983). Some very fascinating early maps of the world were made using woodcuts (Wilson and Wilson, 1976). Although cartography and mapping existed much earlier than the fifteenth century, it is notable that the printing press contributed to the expansion and utilization of maps and the development of the cartographic profession. Two of the most famous cartographers lived during that time, Abraham Ortelius (1598) and Gerard Mercator (1594), and contributed significantly to an understanding of the world's shape and to navigating the world (Karrow, 1993).

One of the more interesting exercises that one can perform with cartography is to ask people to make a map of the area they live in. The entities they choose to include and how they represent them in map form are endless. How people choose to represent their homes can be intriguing to watch. Sometimes referred to as a *topic map,* such a map is based on subjects and is described as follows: "In the most generic sense, a 'subject' is anything whatsoever, regardless of whether it exists or has any other specific characteristics, about which anything whatsoever may be asserted by any means whatsoever" (Biezunski et al., 2000). This is a very open statement and would appear to cover many possibilities. By and large, however, when asked to draw their region, most people think of objects and subjects within the region, then begin to sketch and draw their map. Usually, the most immediate issue is in trying to orient toward north. Then they try to visualize their area. Taken to the limit: "Everything is part of a map" (Thurston, 2001).

More recently, a map has been defined with a view to the digital age and includes concepts such as creativeness and art:

> Cartography: The art, science and technology of making maps, together with their study as scientific documents and works of art. In this context maps may be regarded as including all types of maps, charts, and sections, three dimensional models and globes representing the Earth or any celestial body at any scale (ICA, 1973, p. 1).

> The whole of scientific, technical and artistic activities directed at the creation and use of cartographic products (Bos et al., 1991).

> Cartography ranged from the study of information, collected by "surveyors"—using that word in its widest observational sense—to the final reproduction of maps and charts at any scale, on any subject and by any means (British National Committee for Geography, 1965).

There are many definitions of cartography, but the words *representation, reproduction,* and *symbology* appear continually in these definitions. Throughout history, humans beings have struggled to represent their world, from ancient paintings on cave walls to present-day 4D interactive environments which use high-speed modern computers. A map can be very intricate, detailed, and easy to read and understand, or it can be totally erroneous, providing a false sense of belief, leading to misunderstanding and disenchantment. Some maps are designed with a humorous element to appeal and gain attention, such as The Map of Life (Yale University), which includes such places as "Happy Old Age Hall," "Despair Gulch," and "Bottomless Pit." There are also maps for cyberspace (Dodge, 2002), which depict the interrelatedness of the Internet. Another unique map is the GlobeMap (Universal Marketing Co.), a map shaped like both a globe and a planar map. Fortunately, those using geotechnologies are usually familiar with cartographic principles and strive to produce useful, informative, and interesting maps. There are also those who have embraced cartography to communicate in less than accurate ways (Monmonier, 1996).

An interesting fact about maps is that they are reusable. But there are also a growing number of people interested in collecting maps for various reasons, sometimes because the map is unique or may have had a limited press run, sometimes because a map is historical and represents a specific point in history. Many historical maps are auctioned, usually costing less than $10,000, but in 1979 one collector paid almost £400 for an original Mercator map, a truly valuable map. It was during the eighteenth century near Paris, France, that Giovanni Domenico Cassini accurately measured a degree of latitude and later produced a series of map sheets covering all of France

(Thrower, 1991). At that point, maps became science. Many maps are collected because they are cartographically special. They are aesthetically appealing, the information useful, and they are accurate. The cartographer in such cases is not only notable for cartographic professionalism but also worthy of artistic merit. Needless to say, these types of maps are not common but form the bulk of maps that are collected.

Today, many maps are generated in digital form, often for computer display. They vaporize and sometimes disappear in the digital void, unfortunately lost forever. Some digital maps also represent information in new ways using 3D technologies. Never before have cartographers been so capable of producing maps in so many media and of communicating them around the world so fast. The Internet is providing the means to do that. Digital GIS is providing the means to create maps quickly that previously would have taken months or years to represent. As digital mapping is growing, a commonly asked question is: Will there continue to be a need for hardcopy maps? The answer is clearly, yes. Hardcopy maps are the cheapest to obtain and use and are easily understood, provided that proper cartographic elements such as scale, orientation, and date are provided. A hardcopy map is convenient, universally understood, and often available in numerous languages, something that is not true of many digital maps. Most hardcopy maps are produced primarily for ease of understanding. They are therefore often designed for local regions and frequently generalized. However, they are also produced for very specific applications and purposes, in which case they are used by those who have an understanding of the purposes for which the maps were designed.

3.2 CARTOGRAPHIC PERCEPTION

Cartography is concerned with the representation of reality on a piece of paper or a computer screen—it is visual. To achieve a level of representation that is useful to the viewer, a cartographic product or map must have a few basic requirements. These include the following:

- Symbols (marks that represent reality)
- Scale (to represent extent, size, and distance)
- Legend (identifying the symbols)
- Orientation (usually, a north arrow for orientation)
- Legibility (if a map cannot be read, it cannot be understood)

The reader may wonder why we did not include other elements. The date would have been useful to know, for example, as would an indication of who made the map. A datum is not provided, and that would have helped. Attributes are not included in this list because attributes represent subsets of features that symbols in general could represent. As an example, there may be trees as features, but the map is usable without knowing if the trees (attributes) are maples, oaks, or basswoods. The map is still usable even though lacking values for attributes such as height, diameter, and level of canopy closure. There would be a problem interpreting this map if the symbols were incorrect, such as trees and buildings both being represented by square.

For a fairly simple map of the type that most people would draw if asked to show where they live, as we noted in Section 3.1, a scale would allow distances to be determined, angles measured, and area perceived. A legend would distinguish trees from buildings and other features. North versus south can be determined, and hopefully, the map can be read, but a key to the symbols would clear up any confusion. People drawing such a map might include a few lines for roads, a few squares for blocks, and perhaps a store or other nearby landmark, and would probably add a star at the site of their house. But when asked to draw their region, many people will not include a scale or a north arrow, and sometimes such maps are barely legible. After all, isn't a map supposed to be a representation of reality? What one draws on a map is based on what is perceived and observed. Imagine a map where a home is located. The block is small, but the house appears as large as the block. The perception might be that there is a big house on a small block because that is what is observed and is represented. Without a scale it is difficult to determine if the block is large or the house is large, or the block is small and the house is small. Both are represented symbolically.

Most of us know that individual houses do not cover entire blocks; therefore, our perception is that the house is poorly represented and the map is erroneous. One way of representing information is to categorize information by type and then link symbols to represent those types (Wood, 1972). In the simplest form, symbols such as small houses could represent houses, straight lines would represent roads, and circles might represent manhole covers. Symbology closely associated with realistic objects is termed *mimetic symbology.* The choice of symbols greatly affects viewer perception and can lead to more effective and quicker communication. Thus, the cartogra-

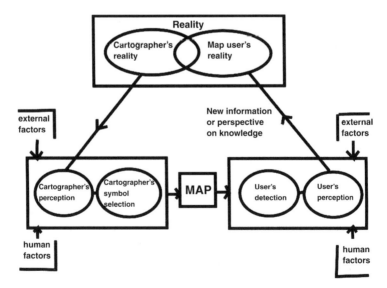

Figure 3.1   Cartographic communication model. (After Wood, 1972; with permission from the British Cartographic Society.)

pher's choice of representation through the use of symbols and the viewer's perception are identified as integral elements in map interpretation (Figure 3.1). Good maps are designed and made for people to understand.

Users of geotechnology are not necessarily thinking of cartographic representation when information is collected with a GPS navigator. But they may be thinking of cartographic representation and symbology when performing GIS analysis and producing a map. The GPS navigator collects vector information in point, line, and polygon formats. Each of these pieces of information can be transformed further using GPS mission planning software. Lines can be reduced to points, or points may be linked to form lines; a series of points forming a line may become a polygon.

3.3 ELIMINATION

The elimination of points poses unique problems for the cartographer. This is required where generalization or a reduction in the level of detail is necessary when moving from large-scale to small-scale representation. Eliminating too many points reduces accurate portrayal. This becomes necessary on small-scale maps to avoid overloading the map with symbology. Two researchers devised a method for eliminating points (Douglas and Peucker, 1973). The *Douglas–Peucker* (pronounced "Poiker") *algorithm* calculates the straight-line distance between two endpoints; then using a threshold distance, it determines the perpendicular distance of points from this line. Points falling below the threshold are eliminated and those above the threshold are kept. In this way, line features are represented while maintaining their characteristic shape. The Douglas–Peucker algorithm is interesting when one thinks about GPS.

Since a GPS can be preprogrammed with a view to navigation as well, it would not be wholly out of the ordinary to program a GPS with the algorithm—collecting and processing points with elimination in mind. It would also be interesting to couple this to GIS, where GPS/GIS data points could be processed automatically. This is theoretical, of course, but shows how integrated geotechnologies could be used to incorporate such algorithms. If one knows that he or she is going to produce a map of specified scale, why waste time collecting unnecessary points which will only be deleted during elimination procedures? During data collection the GPS user is focused on navigation and positioning of points, lines, and polygons; the GPS user may, however, consider scale. Since GPS regularly achieves accuracies of 1 m or less, there will be a need later in a project to represent information at that level of accuracy or scale.

3.4 AGGREGATION

Consider a series of GPS waypoints collected for a city block. Resolution of the information is 1 m and can be represented at 1 m. The points cannot be represented cartographically at 0.5-m resolution but can be generalized to 10, 20, 50, or 100 m or more in scale. A 1-m object cannot readily be seen on a map at various scales, so exaggeration is used; that is, for a point to be seen, it has to be drawn larger than it normally would be according to scale. Consider that a pencil tip is about 1 mm in diameter and that a small-scale map (1:100,000) is available. If the pencil tip were represented on that map, the smallest identifiable unit would be 100 m—right beneath the tip (1 unit = 100,000 of the same unit, or 1 mm = 100,000 mm or 100 m; Table 3.1).

The symbols used to represent features at a scale of 1 m can become quite congested on a small-scale map. In this case the map would probably be useless. Alternatively, at a small scale, the GPS points might be so generalized that they can barely be seen. One way of representing these points at the same resolution that allows for them to be represented is to aggregate individual points into classes. Using this technique, the group of points could form a polygon and be seen more readily. Aggregation is not something that should be taken lightly, however, since it can lead to misinterpretation. If the GPS points are to be represented in small scale, entire blocks might become aggregated and all blocks represented collectively. But if there are five classes of features within a block, each noted as a waypoint, the features must first be classed, since they are distinctly different from each other. For example, imagine a series of city blocks with house heights classed by 10-m increments. Each house is identified with a unique GPS coordinate (geocode) and assigned a value (0–10, 10–20, 20–30 m, etc.).

Breaking data into classes is often referred to as *ranging* or *range grading*. When the values are aggregated by blocks, one block may have 20 houses in one range as compared to another block with two houses in the same range. Aggregating by blocks might not be the best way to go if this difference is to be maintained and represented, which is one of the values of a GIS. Using GIS, individual blocks can be represented or an entire series of blocks may be aggregated, depending on what needs to be represented. The relationship of one feature (variable) as compared to another, more specifically the distribution of one feature to another, is called *thematic mapping*. Thematic maps can be either small scale or large scale. If small scale, features are being compared over larger areas, suggesting a need to maintain large-scale differences, comparing individual blocks by blocks or series of blocks.

The aggregation of spatial information for homogeneous entities does not demand that large-scale differences be maintained, since what applies to one block also applies to another. Even a GPS operator would be happy if all blocks (using this ex-

Table 3.1 Observable units by scale

Scale	Ground Distance (m) / mm on Map	Scale	Ground Distance (m) / mm on Map
1:1,000	1	1:30,000	30
1:2,000	2	1:50,000	50
1:5,000	5	1:60,000	60
1:10,000	10	1:100,000	100
1:12,000	12	1:250,000	250
1:15,000	15	1:500,000	500
1:20,000	20	1:1,000,000	1000

ample) had houses of similar heights that were evenly distributed. They could then locate two houses and assign values to all the rest without even having to collect the data. Their perception then would be that all houses are the same for all blocks, but reality is not this straightforward. Houses vary in height from block to block. This observation requires that consideration be given to representing those relationships accurately. A thematic map is an effective means for making those representations and analyzing those relationships.

One way to begin studying the relationships of blocks to houses is to create separate thematic layers or themes (see Chapter 4). To do this, information related to the blocks, *cadastral information,* could be compiled into one data set, a *cadastral map.* Included would be all the coordinates relating to the blocks, roads, and property boundaries. A series of cadastral maps are referred to as *plans.* You may have seen development plans for a region prepared for those buying property. If the information on the houses were collated, it could form its own thematic layer. This would include the houses and their dimensions and any other information related to them. Such maps are deemed *development or infrastructure maps.* These two themes can now be input to GIS and analyzed with respect to each other. But we might also want to know other things pertaining to the area, such as topography—we wouldn't want to build on a very steep hill and probably wish to avoid low-lying, flood-prone areas.

Using a GPS, (x,y,z) waypoints can be collected for the purposes of building a topographic layer, which we would call a *topographic map.* If the data were input to GIS, a continuous surface model could be generated. A topographic map will have contour lines indicating elevations, and index contours will be assigned numerals, since not all contour lines need to be labeled. Contour lines on a map that appear more closely spaced are indicative of steeper slopes; those farther apart represent less steep or perhaps flat terrain (Figure 3.2). A topographic map can be viewed in 2D or 3D,

Figure 3.2 Contour intervals.

although most often they are viewed in 2D. The figure includes independent GPS waypoints, which have (x,y,z) coordinates that were then used to create contour lines. In practice, the waypoints would not be included, only the contour lines for presentation as a topographic map. A topographic map is represented most accurately in 2D format using an orthographic view. In 3D, particularly without a scale, it is almost impossible to determine slopes and heights as the map rotates and viewpoints vary.

If the topographic data were used to create a theme, the relationship between the contour-line theme and the blocks and/or houses themes could be studied. In Figure 3.3, the house at A is on flat terrain, whereas the house at B is on steep terrain, indicated by more closely spaced contour lines. A GPS has been used to collect all this information, and a GIS was used to generate the contour lines and blocks shown. The map contains a scale and a legend, orientation is noted, and symbols are used. It would appear to be legible. That just about covers the list of requirements listed above, which we indicated is all that is needed to create a map— or does it? Notice that the house at A does not seem to be located within the block; it seems to be partially on the street. This may be due to a number of factors pertaining to GPS and GIS. A GPS will generally achieve ±37-m (at 95 percent) accuracy uncorrected, that is, without differential GPS processing. Perhaps the location of this house was mapped using uncorrected GPS data. Small GPS location differences can affect the topographic map. Alternatively, perhaps the data were collected using the World Geodetic System datum (WGS84), but when the GPS data were transferred to the GIS, they were converted to the North American Datum (NAD83). Areal distances are known to be different for different datums. Possibly the scale has something to do with the

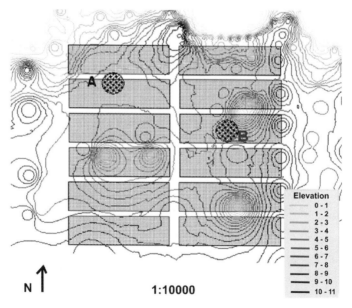

Figure 3.3 Topography error due to GPS errors.

representation. At 1 : 10,000 the smallest identifiable unit is 10 m and depending on whether the data constituted a matrix or vector model (raster or vector), representation of an (x,y) coordinate may not have been achievable.

This raises questions about how the house was geocoded. Did the GPS operator stand on the sidewalk in front of the house, was it geocoded by address according to the plan, or was the center of the building used to geocode the house? All of these are possibilities, and we discuss raster and vector models in more depth in Chapter 4. For now it is enough to realize that map representation using geotechnologies involves many considerations if a map is to be integrated properly. A person viewing the map in Figure 3.3 cannot be quite sure of the status of the house at A and may wonder if the house at B has been placed correctly. Are the blocks correct? Perhaps the entire map should be viewed skeptically.

Thus, it is easily seen that poor representation leads to perceptions that cause doubt as to a map's quality. If the quality of a map is in doubt, surely its usefulness is also in doubt. To clarify this example, let's go back to our original list because it is now clear that it needs a few additions. The date, datum, and inclusion of technologies and methodologies or metadata would be useful. The fact that the datum is not given is a clear sign that it has either been left out on purpose or through confusion (and that may be why the house is in the middle of the street). Datum issues are a major consideration when integrating geotechnologies. So a useful, high-quality map will include:

- Symbols (marks that represent reality)
- Scale (to represent extent, size, and distance)
- Legend (identifying the symbols)
- Orientation (usually, a north arrow for orientation)
- Legibility (if a map cannot be read, it cannot be understood)
- Datum
- Date
- Metadata reference
- Cartographer's name

The datum is very important because the Earth is not round. It is wider at the equator and decreases in size as one moves away from the equator, depending on the datum used. For the WGS datum, the length of $1°$ of latitude and longitude varies in relation to the equator (Table 3.2).

Map projection also affects perception. There are many map projections, and each includes some distortion in its representation (Robinson et al., 1995). GPS navigators include provision for recording the datum but often do not represent that information on the light-emitting diode (LED) using different projections. Because a GPS LED is small, only a small planar format is represented. Consequently, anyone attempting to see a map representation on a GPS over a large area using a specific projection is unable to do so. This does not affect GPS operation, since most users are usually gath-

Table 3.2   Distance variation by latitude and longitude

Latitude (deg)	Kilometers	Miles	Longitude (deg)	Kilometers	Miles
0	110.57	68.71	0	111.32	69.17
10	110.61	68.73	10	109.64	68.13
20	110.70	67.79	20	104.65	65.03
30	110.85	68.88	30	96.49	59.95
40	111.04	68.99	40	85.36	53.06
50	111.23	69.12	50	71.70	44.55
60	111.41	69.23	60	55.80	34.67
70	111.56	69.32	70	38.19	23.73
80	111.66	69.38	80	19.39	12.05
90	111.69	69.40	90	0.00	0.00

ering information for smaller areas where projection is not an issue (with projection differences of only about 1 m for most large-scale work). The projection of GPS data can be handled in mission planning software and GIS software, also discussed in Chapter 5.

3.5   CLASSIFICATION

When collecting GPS data points and features, a need arises to classify them into sub-groups. Classification is an aggregation process that can be based on features, attributes, or values. Some examples of classification are:

- *Width of roads* (in meters): one, two, or four or more lanes
- *Species classification:* such as pine, birch, palm, or spruce
- *Conditional coding:* such as poor, good, or excellent crop health
- *Numerical range:* such as age, heights of trees, or numbers of customers
- *Density:* of, for example, temperature, people, or water quality

The classification method used depends on the variable type, frequency, and distribution. Spatial information can consist of a population with unique entities that may occur regularly and be distributed evenly or unevenly. There may be areas where higher occurrences of entities occur or areas where they occur only a few times. Each of these may be expressed through numerals or may occur as observations with descriptors. Numerical variables with values assigned are easier to classify and query in database form. The inclusion of descriptive annotation requires that thought be given to strings of information and their description.

Classification is easier to accomplish where large jurisdictions and/or professional requirements and standards are in place. Cities, counties, provinces, states, and countries tend to collect spatial information independently, using similar methods and techniques. This allows, for example, census comparisons across a jurisdiction, but comparisons between jurisdictions are more difficult, due to the varying methods and formats utilized in each jurisdiction. Thus, it can be seen that as classification between jurisdictions becomes problematic, the ability to classify decreases. Fortunately, that is beginning to change as data standards change, which in turn is giving rise to national spatial data networks with seamless data capability. Some professions require that spatial data be collected using specific techniques. This is more common in the GPS industry, where data variability arises due to GPS anomalies and operating characteristics.

Nevertheless, the end user of spatial information is usually interested in quantifying differences within populations of variables spatially. There are many ways to do this, but the most popular modes of classification implemented using GIS are:

- *Quantile* (creates a number of classes, each with an equal numbers of features):

$$\frac{\text{number of features in the topic}}{\text{number of classification classes}}$$

- *Equal interval* (classification of integers and real numbers into equal intervals; each class can have a range of values):

$$\frac{\text{maximum value} - \text{minimum value}}{\text{number of classes}}$$

- *Unique value* (creates individual classes for each value; useful for strings, integers, and real numbers):

$$\text{number of unique values} = \text{number of classes}$$

- *Natural breaks* (determines the naturally occurring classes—breaks over the distribution):

$$\text{number of natural breaks} = \frac{\text{number of similar values}}{\text{number of classes}}$$

The selection of which method to use is determined by how the data are to be represented. Classifying one data set using each of the quantile, equal interval, unique value, and natural break methods will result in sets of results: elevation data in Figures 3.4, 3.5, and 3.6. The representation for each of the quantile, natural break, and equal interval methods are shown (Figures 3.4, 3.5, and 3.6). The unique method results in all dots having a different color. Understanding the methods used to classify variables, together with some knowledge of the variables being classified, promotes effective communication of the variables under investigation.

Not all variables occur or appear with clear and distinct borders. Soil samples are

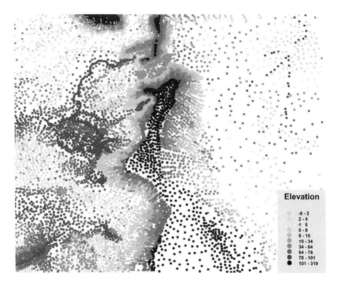

Figure 3.4 Classification by quantile.

a good example. Others, such as air quality, vegetation disease levels, and noise levels, are similar, having no distinct easily measurable boundary. While a GIS is able to analyze spatial and aspatial information, and classes may be determined using one of the foregoing methods, those analysis are based solely on the existing data. Whether a GPS, laser instrument, remotely sensed image, LIDAR, or field observation is used, initially attention must be concerned with, and address, issues relating to

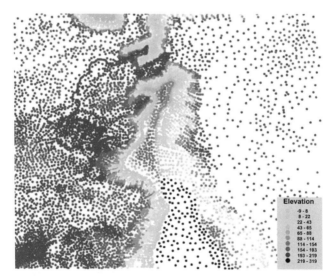

Figure 3.5 Classification by natural breaks.

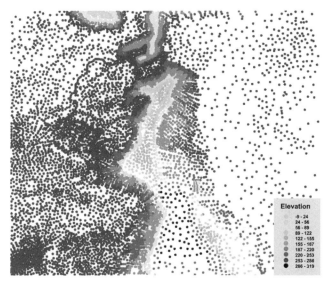

Figure 3.6 Classification by equal intervals.

the collection of information—or whatever most truly represents the phenomenon under study. This implies that knowledge about the phenomenon must accompany the development of suitable methods for the collection of data. This is one of the most widely understated facts in the use of geotechnologies.

Often, users of geotechnologies will collect all sorts of information because the instrument or technology can, without thought to what needs to be collected, produce a sound cartographic map. As an example, consider the use of data loggers. They are often capable of collecting information every second for very long periods of time. These files may grow very large and become burdensome to analyze and use, so much so that they might not even be able to be loaded into a desktop computer, being hundreds of megabytes or more.

Using a 1-second sampling interval, an array of sensors used to measure air temperature could conceivably collect (86,400) values in a day. Even the smallest arrays of sensors in a mountain valley for avalanche studies would collect huge amounts of data using this sampling interval; 10 sensors would collect 864,000 values in one day. Immediately, one must ask why such intensive sampling is needed. But consider the representation of this information: How can it be shown on a single map? It could be produced on 24 maps, one for each hour of the day, but why would anyone want to do that? After viewing three or four maps the human mind sufficiently confused, such as not to be able to determine changes between maps, so the representations would all blend together, resulting in mass confusion.

The 864,000 data points could also be represented using equal interval, natural break, or quantile classification. The ranges of the classes might vary depending on the method being used. It becomes clear very quickly that there is no difficulty collecting data using geotechnologies but that problems arise at the stage of representing the data cartographically. It is not too surprizing, then, that large data sets often get transformed into animations or visualizations, due to their continuous nature. Large data sets are more suitable for showing trends (continuous change), and visualization tools are a means to do that. Due to their continuous nature, large data sets are also more useful than small data sets for modeling purposes, as small sets require estimation of unknown variables because they are not sampled in space or time.

How often have we heard that a model is only as good as the resolution and parameterization of the data being used to drive it? For biological phenomena this is particularly interesting because boundaries are indeterminate, consisting of many discrete individuals forming a larger population. Unwin (1975) describes three phases of landscape evaluation: (1) there must be landscape measurement, an inventory of what actually exists in the landscape; (2) there must be an investigation and measurement of value judgments or preferences in the visual landscape; and (3) must be a landscape evaluation, an assessment of the quality of an objective visual landscape in terms of individual or societal preferences for different landscape types.

A structured method of landscape assessment, linking description, classification, analysis, and evaluation, will provide an integrated framework within which land use management and advice decisions can be debated (Cooper and Murray, 1992). In the area of modeling, research has embraced the study of scale in relation to model accuracy. It is critical in the application of habitat-relationship modeling that the effects of spatial scales are well understood (Karl et al., 2000). The representational tools used in georepresentation are, for the most part, derived primarily from database design and from geometry (Molenaar, 1998). Most GIS applications and associated analyses are in 2D when the primary purpose of the GIS application is to register landscape attributes. The representational phase of two-dimensional spatial data modeling is concerned with the range of geometric structures available to represent discretized configurations (Raper, 2000).

Consequently, it is the database that determines if spatial information can be represented. This is important to know, because using our example of 864,000 data points each day, the management of the database becomes a very real issue—simply in storage alone. If maps were produced to a very small scale, one air temperature could conceivably represent thousands of square kilometers. Some might argue that that is indeed the case now, as it is when viewing most major weather forecasts. These are often constructed from regional data of a widely dispersed monitoring network of climate stations. There are two issues here:

- What is the nature of the phenomenon under study, what is its spatial nature, does it need to be studied continuously, and what is the value of the data in understanding the phenomenon?
- Is a map the preferred means of representing the information, or are alternative technologies available (e.g., visualization)?

These are not simple questions to answer. An understanding of air temperature dynamics would be useful, as outdoor temperatures tend to vary considerably. They also vary by time and direction. It may be useful to know if the purpose of a project is to understand the local relationships of air temperature on flora or rather, to acquire information for regional weather forecasting, in which case it may not matter that there are few data loggers representing large regions (although the more stations contributing to such forecasting, the better). Therefore, even GPS data collection or planning for remotely sensed data tools requires consideration of a few issues:

- What is the nature of the phenomena being measured?
- At what scale will they be represented?
- Do phenomena need to be measured continuously or discontinuously?
- What method of analysis is being considered?

Tessellation is used to divide larger areas into smaller ones. The number of phenomena that occur within each polygon unit derived through tessellation may vary. Tessellation is a way to divide large areas so that, for example, the spatial distribution of a population can be analyzed. A regular tessellation consists of several equal-area polygons, usually represented as triangles, squares, or hexagons (Figure 3.7). The area that these polygons cover may be either large or small, depending on the scale chosen, and the number of sample points or entities occurring within each polygon may vary. Assume that each dot shown in the figure is a GPS waypoint. Different numbers of dots are enclosed in each type of polygon.

Now assume that the scale is reduced to 1 : 10,000 or less. There could then be thousands of waypoints within each type of polygon, but not all polygons would have the same number, and we are talking here of a regularly spaced grid of waypoints. If the sample points were distributed irregularly, then depending on their location, a polygon might have zero, a few, or very many waypoints. This is the same principle that we use when talking about datum in GPS and GIS—thus why they must match. Because each datum covers a different area, if a series of GPS waypoints in 10 different datums were taken and then projected using only one datum in a GIS, the individual points would vary considerably in distance. Although that is not tessellation

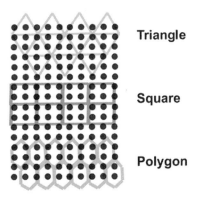

Figure 3.7 Tesselation of GPS points.

per se, the principle of representing area applies. But if tessellation were used, there would be a different number of waypoints per tessellated polygon since they occur in space differently from the way they occur in a GPS datum.

3.6 BOUNDARIES

One area of interest to landscape ecologists is the study of boundaries between ecosystems. In the field, particularly with naturally occurring phenomena, boundaries are sometimes difficult to determine. Historically, ecologists have studied homogeneous regions to characterize and understand ecosystem processes and have avoided the heterogeneous areas between ecosystems (Fortin et al., 2000). The transitions between ecosystems are often indeterminate. To delineate them, a variety of techniques and methods have been used, including shape, preference, quality, and size. A GIS does not render transition zones well, due to the nature of vector and raster topology. This raises a question as to how to construct a map based on topology with sharp boundaries, where vectors or rasters (polygons) do not overlap because of database structure. Instead, they meet at the edges, which in GIS terms is called *adjacency*.

Using GIS, these sharp boundary edges cause viewers to think that water quality is sharply delineated, as are incidence of disease, temperature, light, and a host of other data and phenomena. That, of course, is not usually so. Instead, the boundaries vary and tend to have irregularly shaped borders in 3D space. The *connectivity* between polygons has been determined to be useful in the study of landscape ecology. In an ecological sense, connectivity describes the patchiness of one polygon as compared with the next, including the breaks or continuity between them. Others have concerns about coarse-scale biodiversity mapping and doubt that species richness data collected at coarse scales accurately represent community and ecosystem representation and persistence and have stated (Conroy and Noon, 1996):

- Species abundance distributions and species richness are poor surrogates for community–ecosystem processes, and are scale dependent.
- Species abundance and richness data are unreliable because of unequal and unknown sampling probabilities and species–habitat models of doubtful reliability.
- Mapped species richness data may be inherently resistant to scaling up or scaling down.
- Decision making based on mapped species richness patterns may be sensitive to errors from unreliable data and models, resulting in suboptimal conservation decisions.

In the area of modeling, research has embraced the study of scale with relation to model accuracy. In the application of habitat-relationship modeling, it is critical that the effects of spatial scales be well understood (Karl et al., 2000). Essentially, then, the collection of spatial information for indeterminate boundaries requires consideration of scale. Scale is also a primary consideration for the cartographer seeking to represent spatial and aspatial information. Even if buffer zones are used to indicate transition zones, they are implemented within GIS in such a way that they are proximal to any queried point, line, or polygon—all buffers appearing the same distance again, due to topology. It is for this reason that many researchers consider database structure to be the greatest impediment for translating reality to a map. There are a number of types of predominant spatial patterns, including large patch, small patch, dendritic, rectilinear, checkerboard, and interdigitated (Forman, 1990). Phenomena of a particular type tend to follow most closely one or the other of these patterns. Hydrology is an example of a dendritic pattern; agricultural fields are either rectilinear or checkerboard. We discuss this observation further in Chapter 8 because it leads to some innovative approaches to identifying and segmenting satellite imagery.

Segmentation is a process of aggregating entities into homogeneous regions. For example, we could aggregate all the routes for all the drivers of a delivery company into one large polygon. This could be done for each day, and each polygon would thus vary in shape and show the areal extent of deliveries for the company over a defined time period. It would probably be more useful to know the polygon or territory for each driver. With that information the numbers of deliveries for driver A could be compared with those for the other drivers in a region. This might be useful to show where overlaps in delivery are occurring and to reallocate deliveries to drivers who frequent those areas. Most companies do this simply by telling their drivers not to go outside delineated regions, thus avoiding overlaps. A problem with this approach is that it does not recognize road networks or the traffic flow on them as it varies throughout a day.

A similar situation arises using GIS proximity analysis when new stores are constructed based on concentric circles of larger and larger areas (and population) with the proposed store in the middle, sometimes called *service areas.* How many times have you said: "I am not going to a particular store close by because the road network is poor and access is problematic"? Probably a few times, as most of us have. Thus, when selecting a store to travel to, travel time and preference are factors also, not just spatial distance.

Segmentation can result in the combining of spatial information for a number of reasons. During image analysis, features are classified and combined based on pixel values which range from 0 to 255. Pixel values may be aggregated by color where the ranges have been predetermined—classified. For example, all forest areas would have darker areas and all building areas would have light tones and shades. A map of these differences would clearly distinguish forest from urban areas. We discuss spectral analysis in more depth in Chapter 8, but for now it is important to know that classification is based on the segmentation of pixel values into classes from an image.

If the trucks of our delivery company drivers were equipped with GPS, there would be a continuous record of vehicle movements throughout each driver's region of travel, information that would provide details about which (x,y,z) coordinates had more or less travel time. For example, driver A may detour 4 km for a delivery each day because of a transportation barrier. The barrier could be a river or some other obstruction that exists between the vehicle and a point 50 m away. Why not let the driver on the other side deliver to that location? Travel-time maps are difficult to find because they depend on so many variables, such as the weather, slopes, vehicles, time of day, speed limits, pedestrian flows, and distance itself. Yet effective transportation planning and routing demands that these analyses be performed. For a large fleet of vehicles operating inefficiently with respect to travel times and focused only on service distance, fuel savings alone could be substantial.

The segmentation of population, often called *demographics,* may include location, age, gender, income, tax rate, numbers of children, and employment as variables in any geographic region. Most of these are spatial data and can be expressed by segmentation techniques using choropleth mapping. As an example, the demographic values are averaged by area and represented using class intervals. When only one of these variables is used, for a choropleth classification that is deemed a *univariate map.* When two variables are being expressed in a map (e.g., age and income), we have a *bivariate map.* A *multivariate map* expresses more than two variables within one map (e.g., age, income, and schools). Any combination is possible, depending on the numbers of variables to be represented.

A GIS is ideally suited for analysis between these classifications. The determination of the number of classes to use is not straightforward because populations do not occur homogeneously over successively larger areas. Having said that, maps with more then 10 classes tend to become difficult to understand unless the classes are unique values. Often, classifications are made based on administrative boundaries. The obvious value of administrative boundaries is for studying jurisdictional relationships. These may include districts, wards, regions, cities, provinces, states, or even countries. It is interesting to note that biologists and others interested in natural phenomena tend to shy away from using administrative boundaries since they do not adequately represent biomes and ecoregions.

As the size of the area being represented becomes larger, it may not be entirely correct to suggest that variables occur homogeneously within the areas delineated. Instead, they may vary considerably with respect to numbers or levels of occurrence, particularly as regions grow large (i.e., the rainfall may occur evenly on one block but not 10 blocks). To overcome this, *dasymetric mapping* is used, based on the understanding that entities do not occur in a homogeneous manner, nor do they necessarily conform to administrative boundaries. Instead, boundaries are represented with respect to knowledge about the phenomenon under study. Rainfall, ecology, income, age, housing type, water quality, noise, numbers of vehicles, and other phenomena can be mapped using dasymetric techniques. Ideally, knowledge of the interactions between phenomena and associated entities necessitate that the cartographer determine where and how to represent the information under review. This knowledge may consist of variables which are similar to one another, distributed within the same area, occurring in similar frequency, dependent or nondependent on one another, or may be less important than other occurrences due to their irregular or periodic occurrence. This is interesting from a data collection standpoint because some people work to collect as many data as possible, then process, filter, transform, eliminate, and aggregate those data as needed until they "tell a story" or can be made into a map. Generally speaking, this approach is not the most suitable, for a number of reasons:

- It does not regard the costs of data collection, management, and analysis.
- It results in excessive manipulation of numbers (and elimination).
- It assumes that the data will be useful.
- There is no project goal.
- The wrong type of information may be collected.

Usually, the collection of GPS data, LIDAR, or image-related information is planned with a view to the phenomenon to be studied. From a research perspective, answers may not be known and a hypothesis must be developed and information collected that either supports or does not support the hypothesis. Careful planning is needed as to which types of information must be collected and which analysis techniques used to defend the hypothesis.

Ultimately, a hypothesis can be evaluated based on statistics. Higher probability results in higher degrees of *repeatability*—recurrence. That is, a certain observation will probably occur many more times than the times it does not. If the task is to stand in one place and take GPS sample points continually, it can be expected that for every 100 (x,y) coordinates taken, 95 percent will fall within ± 37 m and 5 percent will exceed that limit. How far those points exceed the limit, and by how much, will vary. Similarly, when a demographic survey is completed, the population distribution can be represented statistically for a given area.

The interesting factor to consider with demographics is that they change. Populations do not usually remain static. This is due in part to numerous factors, including births, mortality, mobility, age, and development. Consequently, a demographic map requires continual updating. This is expensive to do, which is why they are not updated annually. Therefore, the cartographer is faced with the task of portraying demographics accurately for the time they occur. Many applications in the location-based servicing market are heavily influenced by the need for updating data. It is one thing to build the most advanced location-allocation model, but if the data are not updated regularly, that model is at best a guess about who lives where and their preferences, for example, in shopping.

The segmentation of demographic data provides a means to integrate information using univariate, bivariate, or multivariate techniques, breaking the information into classes to represent population spatial relationships. Color ramps may be used to represent the information, and these may be monotonic, in shades of black and white, or they may be in color, as tones of the same shade or unique and distinctly different colors.

3.7 COMMUNITIES

The ability to delineate regions and neighborhoods has expanded to include the concept of *indicators* (Kingsley, 1999), which is sometimes called *regionalism*. When watersheds and population distribution are used to define boundaries, indicators are measurements relating to hydrology or demographics. Using this approach, several identifiable factors are used to determine the relative healthiness of a community and to serve as indicators for establishing relative *community sustainability*. These may include resident employment, job accessibility, income, access to capital, supports for human capital, neighborhood business activity, youth achievement, and education, and may also include mobility factors and cultural and recreational resources, among others. Many of these types of data are intriguing to gather because they can originate within communities or may originate outside communities. Economic indicators are a good example. They may occur very distant from the region under study, influenced by regional or even national economic decisions. Such information is difficult to assimilate. Once integrated, these indicators are used because they provide clues about the healthiness of a community and whether or not that community is growing, deteriorating, or remaining static.

Mapping the numbers of functioning street lamps, sidewalks with cracks, visual perception of waste, overgrown trees, deteriorating fences, noise levels, travel patterns, transportation access, and others utilizes geotechnologies. Both aboveground (remote-sensing aerial photographs) and on-ground technologies (e.g., laser, data loggers, GPS, instrumentation) may be used. Aerial photographs and remote imaging can be used to determine tree growth. On-ground imaging coupled to georeferenced video can also be used to determine building conditions and/or the conditions of sidewalks and roads. Travel patterns by vehicles, noise, and waste conditions can be ascertained with instrumentation. Personal interviews can be allocated to geocoded residences.

Slowly, a database is constructed using these approaches. Thresholds would need to be determined; for example, what distinguishes a poor sidewalk from a good one? Are they assessed by the number of cracks, the number of weeds in the cracks, or the amount of chipping? When mapped, these types of indicators can provide important clues about the changing patterns of residents within a geographic neighborhood. How far do people travel to go to work? Which roads do they use? Is the crime rate related to the level of lighting, and where? An interesting study was completed recently that mapped the occurrence of graffiti. The idea was to correlate vandalism with the occurrence of graffiti. The mapping of indicators raises important questions about how to represent information from varying scales where subjective and objective observations occur. How would you rate graffiti? How can quality of landscaping be measured?

In another community, Beaver County near Edmonton, Canada, residents in five communities have embraced geotechnologies for the development and coordination of activities within their region (Figure 3.8). Their goal is to link the operations of five communities and the county to develop a regional economic development plan. Using a *sustainable community wheel,* they have identified interactions between the communities and how their governments interface. Ultimately, their goal is to develop a regional sustainability plan within a defined bioregion. What is unique about this is that they have apportioned interactions and associated geotechnologies with them to determine the impact of geotechnologies and their associated benefits through the communities, something for which there is little international information. Almost every government and company in the world attempts to determine how spatial tech-

Tofield, Ryley, Beaver County, Holden, Viking

Figure 3.8 Sustainable communities.

nologies result in benefits to their organization. This project does not stop there, however. In the long term, community data generated in real time would be linked to nearby university computing systems, processed and distributed back to the communities. The benefits of such a plan accrue to the university as well as to the communities. By providing real data for learning and research while fulfilling operational needs at the administrative level, they capitalize on high-speed data networks and supercomputing capabilities in the region while providing useful information for teaching and research.

In the area of sustainable communities, 3D visualization is playing an increasingly important role. Several cities are developing 3D city models that are used to communicate and explore communities and cities. Such models are valuable because they can often be viewed over the Internet by individuals and organizations considering moving to, or investing in, the region. They are also useful for community development and presentation of proposed land use changes. A good example of this can be found in New York City, where following the events of September 11, 2001, several building proposals for the site were constructed in mid-2002. During the community involvement phase, several residents commented on the proposals. After viewing the 3D models, several residents opposed the proposed new structures. During those community debates it was found that one of the reasons that each proposal included several large buildings was due to the fact that the land development must result in almost 11 million square feet of office space. This then led to the conclusion, by residents, that the development requirements need to be changed.

Crime mapping is another area where integrated geotechnologies may be used.

Several techniques are used to gather information for the purposes of *profiling* (Turvey, 1999), which involves the collection of information related to behavior and character used to identify suspected criminal behavior patterns. A key element in profiling is time and place (Canter, 1995). GIS can be used to determine potential mobility patterns. Time-of-day analysis could be used for crime scene spatial analysis in the forensics area. Drive times (walk/run) can be calculated using network analysis. Demographics coupled with profiling addresses issues of likelihood. Most of this work is focused on 2D representation, and little information can be found on 3D analysis and representation of criminal activity. Are slopes related to crime? Are building heights related to crime? How can profiling and buildings be represented in 3D? Can the same methodologies used for healthy and sustainable communities be applied to profiling? Are there connectivity issues related to profiling? Where does exploratory visualization fit into the picture? How can we map profiles and compare them thematically? GIS could play a major role in these areas.

Crime mapping has involved VRML (Lodha and Verma, 1999). Visualization tools could be useful for developing crime scene reconstruction and suitable for profiling applications, particularly where the visualizations are coupled tightly to GIS databases. Visualization is useful because it can be used for communication and exploration, and new technologies exist for visualization that can be transported and utilized in remote locations. The primary uses of geotechnologies by law enforcement agencies (Mamalian et al., 1999) include the following:

- Inform officers and investigators of crime scene locations.
- Make resource allocation decisions.
- Evaluate interventions.
- Inform residents about crime activity in their regions.
- Identify repeat calls for service.

As communities begin to evaluate and understand how they function and gather resident perceptions, concepts of mapping are changing. They are becoming localized and large scale. The ability to represent large-scale information effectively requires technologies that can achieve higher resolutions. Because these types of information are represented on a large scale and data are captured continuously, they require higher memory storage and data access capability. Privacy and copyright are also issues that have to be studied. The monitoring of communities will include applications that involve sensor and instrumentation technologies, including digital cameras and laser and sound measurement instruments, which can capture spatial information related to community activity and change.

3.8 CARTOGRAPHIC DISTRIBUTION

As noted previously, the primary cartographic product is the hardcopy map. People relate to such maps quickly, reusing them over and over, and they often need them where they are located. Maps are purchased based on their *usefulness,* implying that all the ingredients mentioned earlier are in the map. Historically, the distribution of cartographic products has been from the cartographer to the user. While hardcopy maps will endure for many years into the future, map distribution and use are changing. This is a function of the convergence of geotechnologies but is also due to the rise in computing availability and associated digital cartographic tools. Anyone can collect GPS data points, buy a satellite image, or acquire LIDAR, and given a few hours (probably more than a few), they could also use a GIS to produce their own maps. Users have become producers. What has made this possible is more direct interaction with the spatial database. Previously, the cartographer maintained the database, producing products as required; indeed, this is still the case with map serving in most instances.

Since a user may download spatial information from a multitude of servers fairly rapidly and import the information into a GIS, the speed at which cartographic products are being used has changed also. The speed of transmission of those products of can, course, result in economic advantages. Geotechnologies have allowed for the increase in speed by which information is gathered and the rate at which cartographic products are produced and distributed. Real-time data flows are contributing increasingly to the explosion in spatial databases and their use. At the same time, *telematics,* the application of geotechnologies for location-based servicing (LBS), is rapidly becoming the largest integrated use of geotechnologies. Each of these is made possible by advances in telecommunications and the increased speeds of the Internet.

3.9 MAP SERVING

Map serving is growing in importance as more users of geotechnologies begin to share their data and cartographic products across the Internet. It could be argued that map serving has been around a long time, given that computers have always been able to transfer cartographic images. GISs evolved from the practice of moving several users (clients) accessing spatial databases located on mainframe computers by invoking commands from remote terminals, which later were processed on the mainframe. As GIS use moved to desktops, increasing numbers of clients began running GIS applications on their own, without necessarily accessing mainframes. This was particularly true given that earlier data sets were small and the analysis did not require a lot of speed and were fairly uncomplicated. However, the sharing of data between clients becomes problematic without interconnecting each client, forming a network.

The coupling of GPS did not really affect desktop GIS until the late 1980s and early 1990s. By then, second- and third-generation GIS software was becoming available, computer processors had increased in speed, and the cost of disk storage was dropping. The Internet itself was not considered "born" until 1983, and it took many years before Internet service providers were readily available, offering low-cost access for public use. The cost of GIS data storage was and remains a major cost for most organizations and individuals, but storage costs have dropped continually in recent years. Since earlier Internet service baud rates transferred at 14.4, 28.8, or maybe 56k, the likelihood of moving large spatial data sets and images was almost nonexistent, with the exception of such facilities as research and educational institutions, which formed the backbone of the Internet. In other words, map serving for general public use was not practical until the middle to late 1990s.

About this time, GIS/GPS technologies and remote sensing began to converge. Applications involved two or more of these technologies together. Prices for these tools dropped, while Internet hardware and telecommunications capability began to increase in speed and capacity. This is, of course, a perspective from North America, where many homes are now connected to the Internet. Such is not the case in Europe, with the exception of Sweden, Norway, and Finland, where higher Internet penetration occurs. The government of Great Britain aims to have 95 percent of U.K. citizens within reach of broadband services by 2006. Many Internet users in North America are surprised to learn that European users often pay for telephone connections by the minute and that there is not the large amount of competition between telecommunication companies as occurs in North America, although that is changing. Consequently, the sharing of large data files in Europe can become expensive, although prices for broadband services and DSL are dropping.

About 600 million people will access the Internet worldwide, spending more than 81 trillion by the end of 2002 according to estimates from International Data Corporation in Farmingham, Massachusetts. It has been estimated that at that time, 55 percent of all North American homes had Internet service, as compared to 38 percent in Europe. About 56 percent of Australians have a home computer, and 37 percent of those have Internet access. North Americans embrace e-commerce and reuse these services about 24 percent of the time, whereas in Europe, consumers remain concerned about security and there is only about a 4 percent reuse of e-commerce services. Clearly, Internet access to information is a function of cost but also of security. This is also true for spatial information. Users of spatial information, particularly those in information technology business applications, are concerned about security of spatial information. Internet users in Germany, France, and Spain continue to voice concern about security aspects of the Internet.

Another factor overlooked in the European market as well as in other regions is the diversity of cultures and languages, which causes there to be a slower shift to Web-based cartography. Very few organizations and small businesses have the capability to translate documents. When spatial information applications include use of the Internet, issues of speed, cost, and language become more important. Developers of Web-based materials are often perplexed when they realize that many users do not sit on Internet backbones or have ADSL services, but instead, connect to the Internet through regular dial-up telephone lines. It is fairly easy to construct a very interesting Web site on a local machine with crystal-clear graphics, but it is another matter to transfer those renderings out across the world to people with dissimilar systems.

Most Internet users are using 800 × 600 resolution monitors with supporting graphic cards. The number of people using resolutions of 1024 × 768 is growing. Designing and building Web mapping applications that exceed this size can lead to problems for many users, in both acquiring and viewing the information. Because many Internet map products are delivered through Hypertext Markup Language (HTML) and more recently, Extended HTML or XML, one approach to dealing with varying user screen size and resolution is to embed instructions in the HTML coding that recognizes screen and resolution size. The simplest approach is to prepare three sets of

HTML pages for all documents. Using this method, Web access is initiated, resolution determined, then the material redirected to the appropriate HTML pages meeting the access resolution information. Alternatively, several organizations construct only one page size for all documents, usually 640 × 480 or 800 × 600, and align the pages to the left. This sometimes results in a large amount of space to the right, called *white space,* for users with larger monitors. Nevertheless, it provides standard-size access and allows for the construction of Web-based HTML pages based on similar dimensions that do not change perspective or orientation.

3.10 INTEROPERABILITY

As consumers and users of spatial data seek to share files with each other, higher levels of interoperability are required—enterprise Web services. Since many individual organizations use different products to gather spatial information and their GIS systems may also utilize different protocols, the ability to integrate the various protocols is a challenge. *Interoperability* refers to the ability to integrate spatial information from one organization or originating system to another (Figure 3.9). The Component Object Model (COM), Common Object Request Broker Architecture (CORBA), and Java technology are the most notable of these technologies. The XML-based interface specifications, first benchmarked in the OGC Web Mapping Testbed (September 1999), provide a powerful, open environment for incorporating multivendor legacy data and software systems into critical Web-deployed applications. The initial specification has been expanded through the Web Mapping Testbed II (late 2000, early 2001) to include the Web MAP Server (WMS), Web Feature Server (WFS), and Web

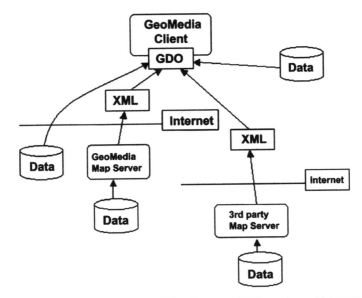

Figure 3.9 Interoperability architecture. (After Guerrero, 2002; courtesy of Intergraph Corporation.)

Coverage Server (WCS) specifications and make use of the Geographic Markup Language (GML). These standard protocols allow disparate systems to interoperate (Guerrero, 2002). GML is an extended version of HTML that includes provision for handling GIS topology such as vectors.

To allow for interoperability, three aspects are considered:

- *OGC Web mapping interfaces.* Common interfaces have been defined to allow clients to connect, query, and display data from various Web servers.
- *OGC Web feature server interfaces.* Web feature servers are capable of delivering data to clients in object form, as opposed to picture form. Web features server interfaces were developed as part of the Web Mapping Testbed 2 during 2000.
- *Web Services Model [World Wide Web Consortium (W3)].* Service provider: entity that creates Web services; service broker: entity that exposes the availability and descriptions of Web services; service requestor: entity that makes use of Web services.

Currently, most cartographic computing is accomplished on local computers—clients. In larger organizations a local area network may be in place, which allows individual computers to communicate with one another through network routers within the organization, forming an *intranet.* The spatial information may exist on one or more computers, with all users potentially having access to spatial information via remote servers. That connection between clients and servers is the basis for map serving. If spatial information is processed locally, this is called *client processing.* If the spatial information is processed on the server, it is termed *remote processing.*

In practice, most networked systems are performing both client and remote processing functions. Those familiar with routers and wiring will know, for example, that if 10 individuals are processing tasks on one remote server, their computing time is shared. The server would cycle through each task, allotting central processing unit time across the combined load. Thus, map servers that are serving very large graphic files can crawl at very slow paces sometimes, particularly where there are several users requesting similar large maps. This has led to different approaches for distributing spatial information. One approach is to continue delivering graphical information. A second approach involves making a call to the server, posing a query for information from the database, retrieving the information, and then constructing a map on the local computer—topological data transfer. Using this approach, the clients continually call the remote servers, acquiring needed spatial data from a spatial data server or warehouse. Whether the spatial information is provided as a database or as a rendered graphic, both are part of the Web services offered within the network.

3.11 TCP/IP

It might be argued that a map server could be a file transfer protocol (FTP) server. FTP users log into a FTP site and download maps and associated files. However, most consider a Web server as being a computer that serves users beyond the immediate organization using Transmission Control Protocol/Internet Protocol (TCP/IP) protocols. The Internet is comprised of a series of networks that use TCP/IP protocol. This allows them to operate as one cooperative, virtual network. Assume that you want to send a map file to a colleague somewhere in the world. The map file is selected, then TCP/IP prepares the data to be sent and received. TCP/IP ensures that a Macintosh, LINUX, or UNIX network can exchange data with a Windows computer, or vice versa. The map that you are sending does not travel to your colleague's computer directly or even in a single continuous stream. Instead, the map gets broken up into smaller separate data packets. The Internet Protocol side of TCP/IP labels each packet with the unique Internet or IP address of your friend's computer. All computers have a unique Internet address.

Since these packets will travel separate routes, some arriving sooner than others, the Transmission Control Protocol side of TCP/IP assigns a sequence number to each packet. These sequence numbers will tell the TCP/IP in your colleague's computer how to reassemble the packets once their computer receives them. This process of TCP/IP exchange is very rapid and takes place in a matter of milliseconds. The packets travel from one router to the next. Each router reads the destination IP address of the packet and decides which path will be fastest. Since the traffic (numbers of packets for several users) on these paths is changing constantly, each packet may be sent a different way. Each point on the Internet that the router uses to send data (step by step in a connect-the-dots fashion) has a variable amount of traffic and variable connection speed, limited primarily by the hardware in place and the sheer volume of packets travelling through that particular location. The packets containing the map information then arrive at the destination, where they are then reassembled in the proper sequence to allow the file to arrive in its original form. This happens every time you access a new map or access an Internet Web page. This is why speed is important and why Web-mapping applications should consider transfer rates.

Every single location (computer, router, name server, hosting account, etc.) on the Internet has at least one unique IP address, which is most commonly represented as four numbers joined by periods, such as 100.100.100.100. These addresses function much like street addresses for the Internet. There are literally thousands of different possible IP addresses, which represent the exact location of a particular site on the Internet and serve to guide any other user to that location. Each location on the Internet might also have one or more domain names that refer to the IP address. In a large enterprise this means that the organization has one address but there are a series of sub-addresses below the main address, called *domains*.

For example, the domain name *welovegis.com* would refer to the location on the Internet, which is currently 100.100.100.100 (which is fictitious for this example). Since it would be impossible for anyone to remember which domain names were matched up to each IP address, the Internet has large computers called *name servers,* which serve as translators. The name servers can look up any domain name and resolve it into the IP address to which it refers. Whenever a user refers to *welovegis.com* or any other domain name, that information is sent to a name server (normally, at the Internet service provider of the user), and the name server returns the matching IP address so that the site can be visited. All of this happens very quickly without user awareness. IP addresses and domain names are not assigned arbitrarily—that would lead to confusion. An application must be filed with the InterNIC, the organization responsible for handling addresses and domain names for each Internet domain name needed. Each domain name must have a matching IP address somewhere on the Internet.

Therefore, when we are "surfing the Web" looking for maps and spatial information, we connect to an Internet service provider (ISP) using the TCP/IP protocol. The ISP has one or more name servers available to you for domain information. You type a domain name (such as *integratedgeotechnology.com*) into the Web browser, or click on a hyperlink from another Web page, and that request is sent at the speed of your modem/ADSL connection to your ISP's name servers. Assuming that the domain name is properly registered with InterNIC, the domain is resolved into an IP address that specifies your true destination on the Internet. The request travels through many routers at varying speeds and pathways until the request arrives at the destination server. If the Web map server is up and running correctly, the destination begins to send information back to your computer in the same manner, until the Web page is visible on the local Web browser.

Hopefully, the map images appear quickly using a Web browser and the viewer can view the map automatically. The computer accessing a server is called a *client,* and the information server is called the *server.* Accordingly, map serving is about the interaction of users and servers (client–servers). Spatial data have gone from being a centralized activity to being a distributed activity among many users. The focus has turned from the GIS application itself to spatial data warehousing and embedding spatial information within other applications, such as spreadsheets, word processors, and other databases. This change has meant increasing focus on the spatial data or the database itself, as well as interface protocols.

The spatial information is used throughout organizations with respect to overall business goals and objectives. Consequently, spatial data servers are replacing stand-alone desktops, which previously housed large data files. These servers are often referred to as *spatial data servers* although they are sometimes called *spatial data warehouses.* They are optimized for concurrent access by many users and provide a wealth of database management functionality.

Access to this information is through local area networks (LANs) and/or the Internet. In some cases, applications are being managed and run on centralized systems. This almost sounds like we have gone full circle, arriving back at the centralized systems of the early 1980s—with a few major differences. Transfer speed, processing speed, and interactivity have increased markedly. Networked computers may be configured together. A LAN couples a file server to two or more networked computers for a working group (i.e., department A). If more working groups are added (i.e., departments B, C, and D), they form a wide-area network (WAN). Each computer within each group would access the same *spatial data server.* Additionally, the data server may be connected outside the organization via a map server. If an application server exists, applications such as GIS can be run on the application server. A good example of this is running Web mapping software on the applications server, which is then served to the world on the Internet via the map server. These specific configurations vary for individual organizations but are similar in many respects for most.

If an application server is used, programs such as Java, GeoVRML, VRML, C++, Visual Basic, and Fortran can be allocated and run on that server. Thus, application servers are noted for having fast processors and are configured with most of the organization's geospatial software. Clients access the server to process spatial data retrieved from the spatial data server and transfer those results within the organization or externally, using the map server for cartographic products. The production of maps for Internet use involves several considerations, particularly an understanding of the viewers' needs from both a technical and a conceptual standpoint:

- Who will be viewing the map?
- What browser type are they using?
- At what speed will the viewers connect to the Internet?
- What do the graphics cards represent, and how do they regenerate?
- How many viewers will access a map server?

Within an organization that proposes to include map serving, a need will exist to have someone who understands both networking and how to maintain and operate a map server. Often, these tasks can be accomplished by members of the existing networking computer group. Alternatively, some organizations opt to use Web hosting services. These consist of third-party computer farms, where dozens of computers are interconnected, storing and processing information. These then relieve organizations from owning and maintaining their own servers.

The OpenGIS Consortium (OGC) standards help to increase the use of remote serving. OGC is an international industry consortium of over 220 companies, government agencies, and universities participating in a consensus process to develop publicly available interface specifications. OGC interface specifications support interoperable solutions that "geoenable" the Web, wireless and location-based services, and mainstream IT, and empower technology developers to make complex spatial information and services accessible and useful with all kinds of applications (OGC, 2001b). Geographic Markup Language (GML) is being developed for the purposes of Internet GIS representation. The goals of GML are (OGC, 2001c) as follows:

- To provide a means of encoding spatial information for both data transport and data storage, especially in a wide-area Internet context
- To be sufficiently extensible to support a wide variety of spatial tasks, from portrayal to analysis
- To establish the foundation for Internet GIS in an incremental and modular fashion and to allow for the efficient encoding of geospatial geometry (e.g., data compression)
- To provide easy-to-understand encodings of spatial information and spatial relationships, including those defined by the OGC Simple Features model
- To be able to separate spatial and nonspatial content from data presentation (graphic or otherwise)
- To permit the easy integration of spatial and nonspatial data, especially for cases in which the nonspatial data are XML encoded

- To be able to readily link spatial (geometric) elements to other spatial or non-spatial elements
- To provide a set of common geographic modeling objects to enable interoperability of independently developed applications

3.12   USEFULNESS

The usefulness of a cartographic product acquired through the Web has received considerable attention recently. In terms of Internet use by country, Sweden ranked at the top in 2000, with 65 percent of the population using the Web, followed by Canada with 60 percent and the United States with 59 percent (Mariano, 2001). Although not all of these Internet users access maps, it has been suggested that the market has reached a plateau—probably not true. The usefulness of information will become the next impetus toward increased Internet use. Cartographic products certainly fall into this market. When designing maps for use on the Web, even experienced cartographers have to adjust their map design habits toward the nature of the Web and what it can offer (Kraak and Brown, 2001). It is not uncommon for users to find that once a link is clicked, a map does not present itself. These are dead-end links and appear for numerous reasons: for example, if the Web browser is not supporting Java, VRML, or another language; or data transfer speed is so low that the browser "times-out," and in rare cases the user's computer logs off automatically from the Internet service provider.

Usefulness implies that the viewer must be able to download or view the intended map being served—the content must be accessible. Maps may be static and presented as single maps or perhaps use multiple representations. A consideration with multiple representations is the size of the average user's monitor. Many people creating maps use large computer screens, whereas the average viewer uses a 14- or 15-inch screen. This means that a small screen can easily be overrun with representations, creating confusion. Even if each window were resized smaller to fit many maps, what often results is that they then become too small to see collectively. The viewer must alternate between screens immediately. This requires continuous tracking and clicking of the computer mouse. One option to avoid this scenario is to buy a larger computer monitor; however, on limited budgets that might not be a first priority. The use of "roll over" or "mouse-over" objects becomes feasible in some applications. When the mouse passes over an object on the screen, it automatically invokes a subset of information. The problem here is, how does one retrace or move backward through objects already clicked? Navigation of the Internet map is thus an extremely important consideration.

Map use is more directly related to the content being offered. This includes how it is being offered and why it is important to be viewed in the first place. It is fairly easy to place dozens of maps into a Web server, but if a viewer must search through them all, it becomes tedious and time consuming and leads to less interaction. Many of the techniques discussed in Chapter 9 apply to map serving as well as visualization. Ideally, a map being delivered over the Web answers questions related to where, what, when, why, and how. The degree to which these questions are answered results in greater interest in the map—more answers, higher use. The use of text should not be avoided when creating Web maps. Well-written text can answer many questions, guiding the user through a Web site toward selection of maps desired to be seen. Text refreshes more quickly than do images, resulting in quicker navigation.

Some maps can be downloaded and in some cases are in a format suitable for direct import to GIS. This has major advantages because the viewer can then work with the information as well as appending other information or deleting as desired. Web map servers can produce themes or layers in some cases. The viewer can query the map and a new map will be created based on the query. These forms of Web mapping have higher levels of interaction, generating higher interest and are more useful. Web maps that include legends, are dated, and include metadata are more useful than those that do not. Some Web sites include games or queries about maps as a means to engage viewers.

The ability to print a Web map is important. Using a hardcopy book or map, the user can see color, but if the viewer does not have a color printer, the map cannot be printed in color and therefore loses a part of its usefulness. Again, this depends on the content. Where Web maps are rich in color and color ramping has been used, those colors cannot be reproduced easily on a black-and-white printer—some blending occurs between classes.

Mobile mapping is being used increasingly for many spatial applications. This involves a GIS, Web map servers, and often the use of handheld portable display devices [personal digital assistants (PDAs)]. How big does a map have to be before it can be interpreted easily? Is a PDA display of a map large enough to be recognized easily? If the map is in color, can it be seen easily in the monotones on a cellular phone or PDA? If the map is large, what does it cost to download a map where network time is paid for by the minute? These are important issues that are related to access to and usefulness of spatial information using these products. Because technology permits an application to be developed, it does not necessarily follow that the application will be useful from the user's or cartographers' standpoint.

Historically, maps have been used for navigation and understanding of relationships between spatial variables. The Internet provides the means to construct and share maps of alternative subject matter and maps that might otherwise never have been produced in hardcopy, due to cost or market. Therefore, Web mapmaking is affecting cartography in new and different ways. The Internet opens the door to the sharing of new content and alternative viewpoints using several different methods.

EXERCISES

3.1. Discuss the elements that a cartographer uses to produce a map, and why they are important.

3.2. Classification is used to identify spatial information for inclusion into groups. Briefly indicate two classification methods and their advantages and disadvantages.

3.3. Two- and three-dimensional surfaces require different approaches with respect to classification. What are some considerations when classifying entities in 2D as compared to 3D?

3.4. How can GPS be used to classify information for field-data-gathering purposes? What approach would you take in performing these classifications, and what are the considerations?

3.5. How does the nature of a phenomenon affect classification?

3.6. Name three types of tessellation. For each of these, discuss how they affect classification.

3.7. Briefly describe the process by which a map is distributed to the Internet. You may choose to include a diagram.

3.8. Map serving allows maps to be transferred to the Internet and viewed. What are some considerations when designing maps for Internet use?

3.9. Define *segmentation* and how it applies to demographics. Provide an example.

3.10. Your job is to delineate boundaries of water quality in a lake. How would you go about acquiring the information and determining those boundaries?

3.11. What is generalization? Discuss generalization with respect to scale.

3.12. Discuss map symbology, indicating what constitutes effective symbolic representation as compared to noneffective symbolic representation.

4

GEOGRAPHIC INFORMATION SYSTEMS

4.1 INTRODUCTION

Over the last decades, spatial data have become more and more important. The reason for this is multifold. First, with the increase in world population, the demand for the well-planned use of available land has increased as well. Such use demands maps and spatial analysis to produce more maps. The second reason for the rising popularity of spatial data and their use is the availability of the new technology called *geographic information systems* (GISs). Since the early 1960s, first researchers and then developers and users converted spatial information into digital form and developed programs that store, manipulate, and display the information.

When asked by someone unfamiliar with GIS for an explanation of their work, many GIS people say something like "I make maps by computer." This is the most visible component of our field but may not be the most important. We are then often asked whether we use "satellite pictures" for that. Such a question often generates a discussion that ends up in a description of the main components of GIS: spatial data gathering (often undertaken by other spatial sciences), databases (perhaps the largest component), visualization, and so on.

It would not be wrong to put databases into the center of a GIS definition: GIS is "a system which uses a spatial database to provide answers to queries of a geographic nature. . . . Since putting spatial data into a computer at great expense for the sole purpose of getting it out again would be pointless, a GIS must allow a variety of manipulations to be carried out, such as sorting, selective retrieval, calculation, and spatial analysis and modeling. We also expect a full range of functions to allow input of data in map form, and cartographic output . . ." (Goodchild, 1985).

This definition is especially attractive because it gives not only a description of the field but also a list of its major activities. But the field has expanded since this definition was written. GIS started out (in the 1960s and 1970s) with mastery of the hardware through the development of analytical and graphical routines. In the 1980s, the systems became larger, commercial, and based on databases that were a combination of the large, generally available (but expensive) business and spatial databases that were developed in-house and served management of the geographical components of the systems.

For an alternative definition, let us start with a statement: "The problem for GIS is that it has the same name for its theory and its tools" (Poiker, 1999). The problem for the public, especially people who purchase systems and hire GIS experts, is that they often think that by buying the system (the tool), they have also bought the theory. We should therefore ask not one but two questions:

- *What is a GIS?* A geographic information system consists of one or several interrelated programs for the capture, editing, transformation, analysis, and presentation of geographically referenced data.
- *What is GIS?* The question "What is GIS?" is very different, and demands an answer that is lengthy and complex, as it would have to touch on all the skills, concepts, and knowledge that comprise the field of GIS within the geography paradigm as well as within the system itself.

The premise that the two questions "What is a GIS?" and "What is GIS?" require different answers relates to the difference between skills and knowledge, or operational skills (system skills) and the nature of spatial data, and also links to the difference between training and education.

- *Training* refers to the acquisition of system or operational skills: It enables the user to work within a specific system through familiarity with that system's commands and operations. Training relates to the "how" of geographic information. Training is demand-fed.
- *Education* is the knowledge and theory behind GIS. Education provides the ability to synthesize and adapt to new environments using life-long learning and experience. Geographic education provides the recipient with an extensive body of theoretical and conceptual spatial knowledge that enables him or her to operate meaningfully and effectively within a GIS. Education is supply-fed.

For the past decade, we have split GIS into these two terms that relate to the foregoing distinction: geographic information systems (GISs) and geographic information science (GISc). Much of the understanding of systems in this chapter has to do with GISc.

The history of GIS is relatively young, having started only in the late 1950s with some surveyors and geographers trying to repeat manual methods by computer. By the late 1960s, as people started to develop the first conceptual components, a number of government agencies and a few universities (mainly in the United States and Canada) were in the process of developing their own system on large mainframes. The first commercial systems were made available in the middle to late 1970s, with the first systems appearing on minicomputers in the early 1980s. This was also the time when the commercial systems grew large, with mainstream databases attached. The 1990s then experienced the appearance of the PC as the carrier of systems. The late 1990s saw a growing number of large-scale data stores with standardization of the data—for data exchange—going hand-in-hand.

Having presented what makes up a GIS, we think it is important to consider what a GIS is not. It is important to remember that a GIS is not simply a computer system for making maps. Nor is a GIS a general-purpose computer graphics system. GISs can produce impressive, professional maps with many impressive symbols, at different scales and projections. But they can do much more than that. In fact, a true GIS never holds maps in the conventional sense but as a database of coordinates or collections of grid cells. From this geographical database we can produce maps as and when required. In addition to producing and storing our map data, a GIS can manipulate them, since the data are stored as a model of the real world rather than as maps in the conventional sense.

What is their similarity to *computer-aided design* (CAD) systems? Computer-aided design is a discipline that developed in parallel with GIS, and the two are sometimes confused. CAD systems also store coordinates and produce maps with them, but a GIS offers a lot more than a CAD system. CAD allows the modeling of entities such as a circuit board or a building. But CAD does not usually require the same volumes of data as GIS, nor does it attempt to model geography.

To simplify the difference, we can say that GIS uses topology, whereas CAD does not, at least in general. Topology provides structure to the data that goes beyond the visual connections of spatial data. GIS does more than simply retrieve existing information; a GIS can produce new information by combining existing information in new ways. This value-added information can be used to help support decision making. The representation of results from a GIS in map form is just one method of output; other methods can be used to suit decision makers.

4.2 GIS TERMINOLOGY

When dealing with spatial information, we usually talk about features in the real world, data when we have turned them into digital form, and objects when we have added structure to the data. However, this distinction is not followed by everybody, and we find ample deviations from this sequence in the literature. Equally, there are different ways of explaining GIS systems and their associated databases. We approach the subject here by viewing three levels of models: geographic space modeling, representation modeling, and logical models. This gives us an overview of how we see geographic space, how we look at its individual components, and how it is structured in the computer.

But before we do this, we should warn those who work with databases outside the GIS sphere that spatial databases are quite different from the standard relational ones. Standard databases, usually in the commercial world, have much simpler structures than ours. They are easily expressed in tables where each row represents a *record,* and each record in a table has the same types of data per column. In other words, tables are not matrices where every single cell in the matrix has the same data type. In a table, we can have columns for such items as, social insurance number (SIN); family name; first name; address; annual income; age; number of children; marital status; and having passed a medical exam or not. In this personal record, probably an employee's record, the first cell holds an integer number; the next three columns are what we call *alphanumeric;* the next is a decimal number; the next two are "integer" numbers again; and the final column contains a binary number ("yes or no"). The SIN is usually also the *key column,* which allows the database user to combine the table with other tables as long as these tables also have the SIN as key.

Standard relational databases do not always behave so regularly. What happens, for example, if a person changes her name? The actual procedure is a little more complicated, but such a case will be dealt with as a special case, an exemption. But if we wanted to accommodate spatial data in a relational table, virtually every record would have to be handled as an exemption. The reason is that spatial units can be defined by one pair of coordinates (a point), two (a straight line), but also by a very large number of pairs (a curved line or an area) (Figure 4.1). However, these points, lines, and polygons usually have data assigned to them that look very much like the records of the relational databases. So how should we deal with these two so different data sets?

Most GISs therefore maintain two databases, one relational, the other a more flexible (but with weaker rules and query functions) spatial database. The two are combined in different ways, as we will see later. It all depends on how we look at things. Our world can be described very differently, depending on whether we see it from the viewpoint of the farmer who plows the fields, the surveyor who measures the landscape, or the GIS operator who puts the data into a database. As we go from the farmer to the GIS operator, we find that we define our surrounding more and more abstractly. This abstraction is very important because it allows us to distance ourselves more from the day-to-day issues. In the following we look at the way the GIS community looks at spatial data.

Point	x	y
1	2	5
2	2	4
3	4	3
4	5	2
5	7	1

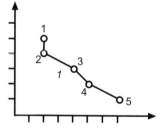

Line	Points
1	1, 2, 3, 4, 5
2	

Figure 4.1   (*x,y*) coordinate route.

4.3 SPATIAL MODELS

We divide space into its parts in different ways, depending on its basic characteristics. We can see space as a jigsaw puzzle of pieces (usually called *regions*) that fit into each other; or we can see space as a continuum, without boundaries, only with different topical intensities, such as mountains but also population, atmospheric pressure, and so on. The first way we can divide space is by using an *object* or *feature-based model*. A spatial object is a portion of the Earth that can be described in a meaningful way. Examples of such features are peaks, churches, and bridges as representatives of point features; streets and boundaries as representatives of line features; and forest stands as representatives of area features, often also *polygons*. More complex examples include the representation of cities by census regions (census tracts) and maps of land use.

The second way we can divide space is by using a *field-based model*. The field view sees the world as a number of variable parameters that are given at every position; examples include topographic terrain and atmospheric pressure maps. The best way to imagine this approach is in terms of a set of surfaces. In the spatial sciences, these surfaces are *contiguous* (without gaps and holes) but not always *continuous* (i.e., the surfaces can have breaks). This distinction does not yet have anything to do with the implementation of spatial information into computer-readable form. It is the way we see the world and how we try to formalize our descriptions.

Spatial Data Primitives

We often talk in GIS of *objects, elements, entities, features, classes, pixels, cells, primitives,* and *spatial,* of course. All these terms are being used in one way or another for the smallest, indivisible component of GIS. In the two data models, they appear in different forms (Figure 4.2).

With *vector data*, points are our basic primitives. All other features are derived from points; only points can be stored as explicit spatial information. Lines are given as ordered series of points, and areas are defined by lists of lines. This means that among other things, a coordinate pair is stored only once and the data set is without redundancies.

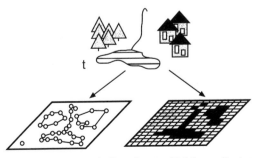

Figure 4.2 Vector (left) and raster (right) coordinates.

In the case of the *raster models,* the question of the smallest building block is defined largely by resolution of the raster. The raster cells or raster points are the atomic elements, which are assumed to be homogeneous throughout their area and allow no further breakdown (at least not in a spatial sense). Lines and areas are composed by a combination of these points or cells.

Representational Models: Vector versus Raster

It is normal that we try to implement features into computer-readable form in a way that duplicates the practical views of our environment. Just think of our standard user interface that is modeled after a typical desk. The fundamental units in conventional databases have definitions and structures that are relatively simple and easy to understand. The objects are usually obvious: persons, merchandise, buildings, parcels, and so on. However, for spatial information, the process of abstraction comes from the weakly understood and goes to the unknown. How do we describe the phenomenon *space* so that we can start with the modeling process?

Let us take as an example the term *landscape,* a term often used in geography but also in other disciplines. The ecologist sees the landscape not so much as an area but as a process of the natural household, the transport of matter and energy, or an intricate set of interdependencies which define an area and let it function as a geosystem. The geologist sees the landscape only for its subsurface, as an indicator of the age and consistency of rocks, and looks for structures and other phenomena hidden under the surface. The surveyor is interested in the determination of exact ownership boundaries. The person doing remote sensing through the use of satellite images is searching for the spectral reflective properties of a square sector of the Earth's surface which can then be interpreted in different land coverages. The farmer sees the landscape as his or her home, understands it through many experiences, and knows about location and capabilities.

These different views of a landscape have one thing in common: They are neither directly visible nor measurable. The ecologist looks at facts, changes, and flows; the farmer collects experiences; the remote senser looks at the area at one very particular point in time and sees average values of reflection of electromagnetic radiation; the surveyor determines the exact location of fixed points; and the geologist derives information on the underground from knowledge of the surface. None of the following objects can be seen directly: wildlife corridors, rock formations, the boundaries of a parcel, the extension of a pixel from a satellite image, and the patch of grass that has to be cut more frequently than other patches. The spatial structures of all five views and their constituent elements are very different from each other. Is it possible to bring those together under a common denominator that will enable us to define them in a uniform conceptual schema?

Layers, Coverages, and Variables

The segmentation of our surroundings into compartments has a long history. Geography, for example, has divided geographic knowledge based on both regional and thematic aspects. Except for a short period in the discipline, the latter has prevailed.

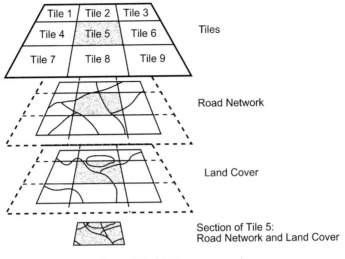

Tile 1 Tile 2 Tile 3
Tile 4 Tile 5 Tile 6
Tile 7 Tile 8 Tile 9 Tiles

Road Network

Land Cover

Section of Tile 5:
Road Network and Land Cover

Figure 4.3 Multilayer approach.

Traditionally, the layer model dates back to the analog origins of cartographic analysis in landscape architecture and landscape planning. Manual overlay has been with us for some time and has been popularized by Ian McHarg (1971). In cartographic production, the multilayer approach (Figure 4.3) has been used for several decades.

Thus, the information layer has become the container of a spatial variable. For example, in the environmental sciences we often distinguish between *physical layers,* such as height above sea level, slope, soil type, lithology, and vegetation, and *cultural layers,* such as land ownership, land price, distance to centers, and distance to roads. A layer could be considered the spatial distribution of one single feature and its storage in digital form. This leads to a vertical organization of a spatial database. Each new entry in such a database is a new layer. The layers are generally assembled independent of each other.

Vector Model

Vector geometry is central to geographic thinking. We talk about points, lines, and areas, not about pixels and cells. Therefore, the vector model in GIS is much closer than the raster model to our thinking about spatial information. (Figure 4.4). The position of the spatial objects is given through coordinates within a reference system, and the common basis is the coordinate pair or tuple (x,y or x,y,z). A single vector defines the position of a point object, an ordered list of vectors defines that of a line, and a closed trail of points describes an area.

Vector coordinates are usually given within a Cartesian (rectangular) coordinate system in the form of absolute coordinates of single points or lists of points for the description of lines. Each spatial object has to be identified with a label for inclusion in a database. The purpose of linking an object to a particular label is to be able to tie attributes to the object that describe properties of the object. A polygon that repre-

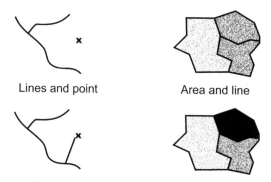

Figure 4.4   Vector versus raster geometry.

sents a parcel, for example, could have as attributes the owner's name and address, the lot size, and some information on buildings on the lot. These attributes create one or several records in the database, and the label represents the key for the link between spatial and attribute data. This *dual structure,* also called the *georelational database,* is typical of the vector data model and is implemented in many graphic and GIS systems.

Unstructured Objects

A major distinction in vector-oriented GIS separates structured from unstructured approaches. *Unstructured objects* are nontopologically organized data sets which have been used especially during the early years of GISs and still are used frequently in CAD and mapping systems, where they play a significant role. In the GIS literature, this approach is often called *spaghetti data* because the individual objects are situated without relation to each other, might overlap, have gaps, and so on. Any inclusion or deletion of spatial objects is trivial and does not need control through the already existing objects. Examples of unstructured spatial data include tracking animals with radio transmitters, and catalogs for aerial photographs, georeferenced literature indices, and so on. In these cases, a forced topology and planar singularity can be a hindrance rather than a help. The region function in ArcInfo, but even more so the shape files in ArcView, are a response to this need.

Building Blocks: Geometric Primitives

All graphic environments that are based on the vector model have primitive objects that are combined to higher objects. Just think of the average vector-based drafting program, which has a set of primitives in its toolbox (line, square, circle or ellipse, arc, and polygon). Similar to GIS, we use the primitives point, line, and area.

Dimensionality   The three GIS primitives represent different dimensions: points are zero-dimensional, lines are one-dimensional, and areas are two-dimensional. When these primitives are combined, then as a rule of thumb, the dimensionality jumps by one. For example, a series of zero-dimensional points builds a two-dimensional line. When primitives of different dimensionalities are combined, the dimensionality of the higher-dimension object is usually maintained. Looking from the other direction, one can say that objects of a certain dimensionality are composed of objects of the next-lower dimensionality: Areas are described by lines; lines consist of a series of points. Further, one can say that the dimensionality of objects is always smaller or equal to the dimension of the space in which the objects are situated: If we operate in a one-dimensional space, along lines, a time axis, and so on, the object in this space can only be points and lines or zero- or one-dimensional. In a planar space, the flatland of cartography, and many data models, we can have only zero- to two-dimensional objects.

Points   Points as zero-dimensional objects are the simplest building blocks of spatial data. Since there are no zero-dimensional entities in the real world, points are usually reduced representations of real linear or areal entities of relatively small (in relationship to the scale) spatial extension. In addition, points are spatial components of lines (vertices, nodes) (Figure 4.5). Examples of points as representations of the real multidimensional entities are topographic elements such as churches, bridges, houses, and power poles. Examples of points as samples or constructs are centroids of areas, addresses, and geodetic monuments. The storage of points is usually in the form of tables with fixed-length records in the form ID (x, y) or ID (x, y, z).

Lines   After points, lines are the most important building blocks in the modeling of spatial information. Lines can represent linear elements as well as entities that are modeled by lines as the borders between areal objects. Examples of linear objects are traffic lines, rivers, contours, and power lines; examples of boundaries are administrative boundaries, coastlines, parcel boundaries, and corridor limits.

Lines are the set of all points that represent the exact course of a line between two endpoints. Since theoretically, any number of points can be positioned between two endpoints, we could get an infinite number of points for a complete description of any line, even a short one. Since any infinite number of units is too much to store, the points selected are representative for the course of the line. Intuitively, we will be

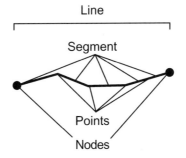

Figure 4.5   Lines, segments, points, and nodes.

looking for characteristic points along the line, points with high information content. These are starting and ending points, intersections of lines and points where there is a marked change of direction in the line.

Thus, lines are *discretized,* that is, the infinite number of points on a continuous line is represented by an ordered sequence of single points. These characteristic points for the representation of a line imply that the line is being stored in a particular range of scale. Discretization excludes a scale-independent storage. Take the case of a natural line (e.g., a coastline). If we look at the line in successively larger scales (with successively more detail), we discover successively more curves in the line.

Simple Polygons   Since the vector model stores spatial items only as zero-dimensional objects or points, areas cannot be stored directly. With surrounding lines, or polygons as they are usually called in GIS, we have the ability to store them implicitly as ordered points. Areas are usually defined by their circumferences (i.e., their boundaries). Polygon closure (Figure 4.6) can therefore be defined in two ways:

- *Implicit:* with identical start and end
- *Explicit:* with a rule that says that the last point is to be connected to the starting point

The explicit case is usually safer since one does not have to repeat a coordinate precisely when digitizing. The function is called *snapping.* It is important for most GIS functions that the start and end of a polygon boundary are identical down to the

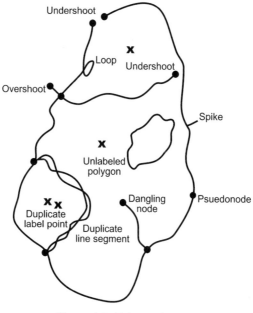

Figure 4.6   Polygon closure.

lowest decimal. Filling areas with color, computing the area, and so on, would create errors, sometimes very irritating ones.

Complex Polygons   Not all polygons can be identified with the conditions described above. We need a few more definitions to account for some deviations from the simple norm. First, we distinguish between convex and nonconvex polygons. Convex polygons have no indentations; that is, every point within the polygon and on its rim can be connected to any other point of the same type without crossing the boundaries of the polygon (Figure 4.7). Such a convex polygon simplifies many of the processes that are applied to polygons, such as hachuring and calculation of the circumference. The nonconvex polygon can have a variety of shapes. It can be concave, have islands (polygons within polygons; Figure 4.8), and so on. Another complexity is the fact that a polygon can consist of several islands (i.e., closed polygons). The most frequent case is the containment of an island within an island.

Topologic Data Structures

Topology is a branch of mathematics that is concerned with spatial properties of discrete objects that remain invariant when distorted. These properties are the topological relations in GIS. The example of rubber sheeting is often used: No matter how much you pull and push a rubber sheet (twisting is not allowed), topological relations between pairs of spatial objects are not disturbed. Topological relations are neighborhood, inclusion, connectivity, and so on. One often distinguishes topology in nets (*network topology*) from topology between areas (*polygon topology*). The creation of topology in spatial data sets is by the setup of tables that contain the relations, usually in the form of pointers (keys, reference labels).

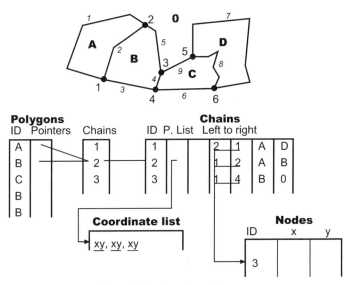

Figure 4.7 Complex polygons.

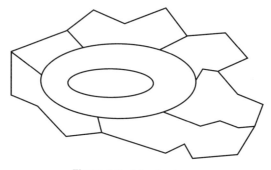

Figure 4.8   Island polygon.

Topologic data structures are often considered the main differentiating feature for GIS, especially with respect to CAD and computer graphics. This is easy to understand: GIS data are some of the largest and most varied data sets in the computer industry, and without a very firm structure, constant changes cannot be incorporated without the danger of making the database ineffective. Furthermore, the type of analysis that is often performed with GIS is not possible without the types of structures that are now available with commercial GIS.

Even though the implementation of topological structures varies from system to system—with the accompanying variation in functionality—some characteristics are basic to all of them:

- Minimization of redundancy (by reusing elements)
- Hierarchical organization with pointers from complex objects to primitives
- Determination of topological properties (contiguity, connectivity, enclosure) from geometrical properties (coordinates)

Topology in Lines and Networks   Topological structures are interesting not only for areal extensions but also for the arrangement of linear elements to networks (Figure 4.9). As a matter of fact, mathematical topology had its origin in the graph theoretical methods of network analysis. Graph theory employs two basic building blocks:

- Nodes
- Edges

A *planar graph* (the most frequent type of graph and the only one that is of interest for us) is a graph lying in a plane or projected into the plane. It is also a structured set of edges (segments, lines, bound by nodes) that are connected at nodes. Edges cannot intersect each other, and if they do, they have to be corrected by making the intersection a new node. Edges can be directed or not directed, and usually carry some attribute information such as transit capacity or flow. Nodes can also have attributes, but their main function is to create a link between edges.

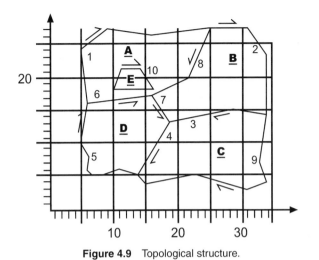

Figure 4.9 Topological structure.

In GISs, network topology is derived from coordinate geometry and stored as part of the latter. In addition to the standard attributes of segments, network edges carry more attributes of topological significance: If the graph is directed, one can distinguish between starting and ending nodes or sources and destinations.

Examples of Topological Structures For the vector case at least, topology is usually derived from coordinates and stored explicitly in tables. The tables have to be updated every time one of the objects in the database is changed (e.g., if a line is deleted). Each time, the topological information becomes obsolete and has to be rebuilt, not just for the object in question but in the neighborhood of the object at least.

Routes Routes are complex objects that are composed by segments of a network, such as a street network. Examples are bus routes, routes for international couriers, garbage collection routes, and snow removal routes. In all these cases, a route is a sequence of network segments that is a unit with respect to scheduling, changes along the route, transport capacity, and so on. An attribute can belong to the entire route (time distance, capacity) or to individual segments (traffic volume, number of lanes, permissible speed) and can even change within a segment. Attributes can be valid for several routes (commuting prices), and individual segments can belong to several routes (bus lines).

Regions In the GIS context, regions are groupings of basic areas—often, administrative areas—into higher-order regions. This is done when basic spatial units such as communities are aggregated into different classes of regions with different objectives. Examples are counties, commuting regions, tourist regions, and planning districts. Each region has its own set of attributes but also needs access to external data such as area and population which have to be aggregated from the constituent communities. To avoid redundancy in the basic attributes as well as in the spatial definition of the regions, regions are treated as complex objects.

Raster Models

Actually, the heading above should be "models of regular and irregular partitions of space," but since the raster is the partition used most frequently, we use this much more recognizable heading. Common to all these approaches is that space is divided into small, mostly regular cells which are considered to be homogeneous and cannot be divided into smaller parts. These divisions are usually called *tessellations*.

Raster Point versus Raster Area   When setting up a raster system, one of the basic concerns is whether the phenomenon in question is considered to be spatially discrete or continuous. In the case of a discrete phenomenon, the tessellation is exhaustive, covering the entire surface (in two dimensions), whereas in a continuum, a unit has to be considered as a point sample (zero-dimensional).

To take the simplest example of a square raster, a discrete phenomenon is divided into raster cells and the complete surface is defined by a grid of cells, whereas with continua, the totality is divided into grid points and we talk about a point grid as the structure. This distinction is important: *Raster cells are often averages or dominant values of the surface, grid points are direct observations and are true only in the center of the cell*. It is possible to interpolate between grid points. Interpolation does not make sense with grid cells. Cells allow only descriptive statistics (complete description of the cell), whereas point distributions also allow sampling (selective description of the cell).

Raster Coordinates   The spatial reference system for rectangular rasters is geometrically very simple. The overall position of the raster is given with respect to a global reference system, whereas the position of the individual raster cells or points is given with respect to the origin of the raster. Contrary to a vector system, where each element is given a coordinate, in a raster system only the attribute information (value of each cell) is stored explicitly. For spatial information, only four pieces of information have to be stored explicitly in order to identify the entire raster:

- The coordinates of the origin of the raster
- The orientation (usually, north)
- The size of the cell (two values for a rectangle, one for a square)
- The dimensions (number of rows and columns)

This information is usually stored as a header to the raster matrix or in a descriptive file (metadata). Other important information is known by conventions without having an explicit place anywhere in the raster files: the position of the origin, usually the top left (in remote sensing systems) or the bottom left [in systems based on mathematical (matrix) notations], the sequence of raster values (row by row, starting with the top row), and so on. These conventions can create difficulties, especially when data sets are exchanged among professional groups or from country to country.

Raster Attributes   In vector systems, there is a dual data model with geometry at one side and attributes at the other. This is not the case in raster systems. In raster systems we have attribute data in the form of the values in the raster cells, whereas the spatial information is given implicitly by the position in the raster matrix and through the global reference of the matrix. Raster attributes are often given in classes or categories (e.g., soil type, vegetation class, etc.) or spatial objects (e.g., parcels, communities, etc.). In this case the classes in the raster can be combined in the legend with a table of other values.

Hierarchical Rasters: Quadtrees

When dealing with spatial data, the density of information varies. We could therefore do with a very coarse grid in some areas, but we need more detail for others. In standard raster systems, the area of highest change determines the grid density. But working at this level is not always economical. Starting with a coarse grid and going into successive refinements when necessary is the characteristic of hierarchical structures, which supply compromises in storage volume, processing effort, and spatial resolution. One of these techniques is called *quadtree*. We start with the entire screen or rectangular area and test whether it is homogeneous. If it is not, we divide it into four equal parts and apply the test again for each parts. This recursive procedure ends either when we have a homogeneous cell or when we have reached an individual pixel.

Because of the hierarchical division of space, a tree structure is useful for the storage of this approach, and since the subdivision always breaks into quadrants, we call the trees quadtrees (Figure 4.10). At the ends of a quadtree, *quad leafs* carry the attributes of the raster. The top of the tree is called the *root,* indicating that we are talking here of an inverted tree. For the purpose of giving the tree coordinates, we also use a hierarchical system so that we can read the level off the numbers. The levels are

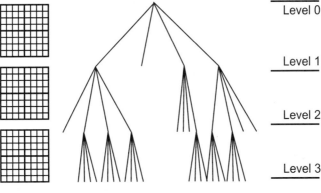

Figure 4.10   Quadtree division.

called *quad levels*. Level 0 is the root, the undivided area; level 1 has four areas; level 2 has 16; and so on. In principle, there is no limit to the number of levels, but the binary storage of the tree suggests limits of 15 or 31 for addresses with two or four bytes. But even with 15 levels, we can achieve a considerable resolution, getting a resolution of 32,768 linear cells. In other words, at this level we can work with an area 32 km square at a resolution of 1 m. At level 31 we can describe the world at a resolution finer than 1 mm.

A quadtree can be built from either vector or raster data. Every distribution of objects that does not match the lowest quadtree level exactly has to be sampled at the density of the level. A quadtree database has the following basic characteristics:

- The quadtree has to be based on a square working area.
- The resolution always increases by a factor of 2.
- An adjustment of the resolution to differentiated structures is possible locally.
- Significant savings in storage can be had with large homogeneous areas.
- Access to individual cells is very fast.
- The address is stored in a single value, not in a coordinate tuple.

Thiessen Polygons

We have used the term *tessellation* before. The leading regular tessellation is the rectangular raster and there is really nothing else for all practical purposes. Thiessen polygons are the most important irregular tessellation and structuring technique. Starting with an arbitrary distribution of points, a *Thiessen polygon* is the area in which every place is closer to one particular point than to any other point in the set (Figure 4.11). Every point in the set gets assigned to exactly the area in which it is the

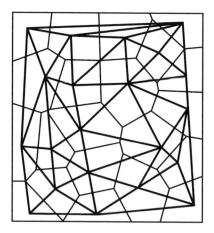

Figure 4.11   Thiessen polygons.

closest neighbor to all possible samples. The Thiessen polygon is thus an assignment of space to points on the basis of shortest-distance criteria.

Thiessen polygons have had a variety of applications, starting with the assignment of areas of precipitation in meteorology (Thiessen was a South African meteorologist) to its use in interpolation of continuous surfaces, the determination of boundaries for nominal regions from point samples (e.g., soil regions), and so on. Of course, if the points are arranged in a regular pattern (e.g., triangular, square, hexagonal), the resulting Thiessen polygons are identical to the regular cells mentioned above. The number of neighbors is most frequently six but can be anything from three to (theoretically) infinity. If you have worked with GPS trajectories and remote sensing data, you will have noticed that they work with totally different types of data. The differences are summarized in Table 4.1.

Surfaces and GIS

Whereas we have concerned ourselves so far mainly with discrete spatial objects, this section is directed to a discussion of continua: Within a study area, a phenomenon is defined continuously and can have any value at any place. There are no sharp boundaries or breaks; thus interpolation is the method used to find values between data points. The most frequently represented type of surface is that of terrain. We give

Table 4.1 Vector versus raster characteristics

Criterion	Vector	Raster
Data type	Vector data are based on precise points. The coordinate systems are either geographic (longitude, latitude) or Cartesian.	A raster consists of a matrix of regularly arranged rectangular cells. Cells are areas, but rasters can also be identified by the centers of the cells.
Point	A point is given in geographic or Cartesian coordinates.	A point is given as an index in rows and columns. Cartesian location can be computed by adding the width and height of a cell, times the index, to the origin.
Line	A line consists of a series of points. The start and the end are often given the term *nodes*.	A line consists of a continuous chain of cells.
Area	An area is a group of lines that create a closed set of boundaries. Areas are usually called *polygons*.	An area is a contiguous set of cells, usually of identical color.
Representation	A point (no area) represents an elementary unit.	A cell represents an elementary unit.
Attribute	Any number of attributes can be associated with an object, usually in a relational database.	One layer (coverage) represents one attribute. There are as many layers as there are attributes.

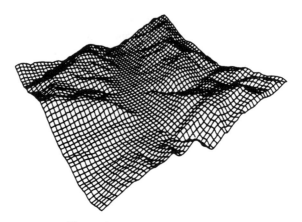

Figure 4.12 Continuous surface.

these surfaces the name *digital terrain models* (DTMs; Figure 4.12). These days, most mapping surveys have complete coverage of their jurisdiction. But there are other uses for surfaces:

- Groundwater levels
- Thematic surfaces (e.g., precipitation, mean temperature)
- Socioeconomic fields (e.g., population potential)

Most of these examples differ from terrain in one point: They are smooth or continuously differentiable (not continuous; they all are), whereas terrain has breaks. This is a fundamental difference and has an influence on all techniques that are used for surface storage, analysis, and presentation. In the following we use terrain as the default surface type and treat smooth surfaces as a special case. However, the term *digital terrain models* is sometimes also used for smooth surfaces, so beware.

Terrain Burrough (1986) presents the following examples for the use of terrain models:

- Cut-and-fill calculations to estimate, for example, the amount of soil and rock to be removed during road construction.
- Three-dimensional display of terrain to assist with the presentation of information. This is particularly useful when it is impossible to visit the region of interest.
- Visibility analysis to determine what features can be observed from a given point in the landscape. Environmental impact projects, landscape planning, and military applications find such analysis of particular benefit.
- Identification of, for example, routes and locations that may be subject to potential hazards, such as landslides or floods.

- Comparison of terrain to assess its suitability for a variety of uses (e.g., agriculture, recreation, mining).
- Production of slope and aspect maps for inclusion in a variety of physical and socioeconomic models.
- As a background for the display of other information, in particular, land cover data such as forestry, land use, and vegetation derived from air photography or satellite images.
- Generation of image simulations of new developments, for example, a ski resort or new forest plantation.
- Representing other surfaces, such as pollution and temperature.

The most common products to be derived from a digital terrain model include:

- Block diagrams, profiles, and horizons
- Contour maps
- Line-of-sight maps
- Slope, convexity, concavity, and aspect maps
- Drainage networks and watersheds

Are digital terrain models three-dimensional? Some writers would claim that they are. But one of the assumptions in the DTM community is that they are *single-valued,* which means that for any combination of x and y, there is only one value of z; in simpler terms, there are no cliffs or overhangs. If nature confronts us with one of these rare occurrences, we either ignore them or deal with them outside the system. This way, the developers do not have to deal with such very special cases; and this means an enormous saving in programming time, storage space, and user headaches. We therefore often say that DTMs are not 3- but 2.5-dimensional.

What if people have to work with *multiple-valued* surfaces? There is a straight way, but it goes through computer graphics. Computer graphics has dealt with three-dimensional objects (usually called *solids*) for years and become very good at it. The tragedy for the GIS people who have a need for their methods is that they have to start learning all over again. There doesn't seem to be any program that links a GIS database with a three-dimensional display program.

Triangulated Irregular Networks

Triangular irregular networks (TINs) represent an alternative to the raster approach. They have been accepted widely by software developers and are being used for many purposes, including contouring and other displays and the computation of volumes. Raster DTMs are usually developed by interpolating irregular data points to a regular grid. TIN networks, on the other hand, usually come directly from the raw data and are triangulated to the network. The original TIN prototypes allowed the selection of significant points within the triangulation algorithm (Figure 4.13).

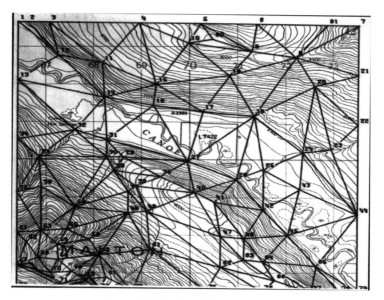

Figure 4.13 Original TIN prototype.

TIN data points are—again usually—chosen to be of maximum information, also called *surface-specific*. In other words, the choice is nonrandom with respect to the surface, preferring extreme points and lines such as peaks, passes, ridges, and valleys. Since terrain is usually not very smooth, the result is a better approximation of terrain, one that can be replicated only by a much denser regular grid. As will be argued later, one TIN point for every 100 grid points is a reasonable ratio, maintaining the quality of representation. This advantage over the raster is bought by a data structure that is much more complex and voluminous per point. Also, all procedures pertaining to the TIN structure are more time consuming because the geometry is more involved. The basic components of the TIN system are as follows:

- Nodes
- Edges
- Triangular faces

The triangle is a very robust spatial object. All other areal units can be divided into triangles, and any surface, however intricate, can be completely covered by triangles. Also, the triangle is the only spatial unit that always represents a plane surface. A plane polygon with more than three corners is an exception, and it would be difficult to guarantee such planarity through the lifetime of a database.

Some advantages of TIN are the following:

- TIN can easily be integrated into vector structures.
- The variability of terrain (flat versus mountainous) can easily be accommodated by using different densities of nodes.

- Breaks are easily represented and not smoothed as in raster systems.
- With similar quality of representation, the storage needs are usually lower.
- Humanly modified terrain, which usually has more breaks than untouched terrain, can naturally be incorporated.
- TIN gives significant advantages to large-scale tasks such as cut-and-fill and exact contours.

There are some critical decisions that have to be made when constructing a TIN:

- *Which nodes to select.* When a large number of points is available—on aerial photographs, contour maps, or from dense regular grids that have to be generalized—the points of high information have to be identified first.
- *How to connect the selected nodes.* Usually, this is taken care of automatically by a triangulation program, which means that the decision is thrown back to the choice of nodes. (A rough rule: If break lines are to be maintained, select the points along the breaks that are denser than in other areas.)
- *How to represent the area within triangles.* Usually, it is treated as plane, but if sharp changes at edges are to be avoided, some mathematical smoothing function has to be applied which guarantees continuity across edges.

Triangulation of a TIN   Triangulation is a process that connects irregularly or regularly distributed points so that none of the connections intersect and the links build triangles that cover the study area completely. Triangulation is generally performed by computer; the manual method is very time consuming. The manual method also rarely produces optimal networks (i.e., sets of triangles that fulfill certain optimality criteria). The most frequent criterion is that of the *Delauney triangulation,* which minimizes the sum of the length of the edges of the triangular network. The Delaunay triangulation is the dual of the Thiessen polygon (i.e., the one intersects the other at right angles). The construction of Thiessen polygons was discussed earlier. If the criteria are followed (for which there are different approaches developed), the points above are connected as shown in Figure 4.14.

Breaks   Breaks are everywhere, in natural landscapes as well as in the humanly modified terrain; and very often, we orient ourselves on these breaks. A map with smooth contours everywhere makes recognition very difficult. Mapping breaks is therefore very important, but it is not easy when the rest of the contours are supposed to be smooth. One usually emulates sharpness by adding an extra set of points. Not that the problem is absolutely insurmountable, but the GIS vendors have not bent over backward to oblige. When digitizing terrain, breaks are often relatively straight and the temptation is great to jump along it in long steps. This will have some curious effects when the points are subsequently triangulated. Valleys show dams and ridges show passes, because points to the side of the breaks cut across them, above the valleys and below the ridges. There are two approaches that can be taken:

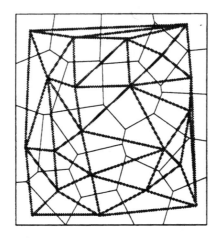

Figure 4.14   Delaunay triangulation: dual of the Thiessen polygon.

- Digitize breaks closer together than other lines. When ridges and channels are digitized, the thumb rule is to make the distance between points a little closer than the distance to the closest neighboring break.
- Record breaks as an additional line and fit it through the network after triangulation. Wherever a break intersects a triangular edge, create an extra point and interpolate its height from the break line (or a weighted average of the two intersecting lines), which gives the breaks realism.

TIN Data Structure   There are two alternatives for storing TIN databases:

- Triangles
- Nodes with pointers to neighbors

For the triangle-oriented structure, we need the following components:

- ID of the triangle
- Coordinates of the three nodes
- IDs of the three neighboring triangles

Since as we noted earlier, a node has six neighbors on average, it is more efficient (i.e., space saving) to keep a separate table with the coordinates of the individual nodes and keep only the node IDs in the triangle tables. In Table 4.2 we show the difference between the TIN and the raster structures.

Metadata

Many organizations collect spatial data. For example, federal governments develop base maps of the country, state and provincial governments collect land use and land

Table 4.2 TIN versus raster characteristics

Criterion	TIN	Raster
Data points	Adaptive to variation of surface	Fixed by orientation of raster
Data point density	Low	High
Redundancy of information	Low	High
Original data	Primary or from maps	Remote sensing, etc.
Direct representation of points and lines	Yes	No
Incorporation of breaklines	Yes	No
Irregular boundaries or holes	Deviation from common	Easy, through masking
Change of detail	Easy, by addition of points	Complicated through global refinement of grid
Storage space per point	50 bytes and higher	2 – 4 bytes
Processing cost for analysis	Medium	Low
Visualization	Resampling to profiles	Direct

cover data, municipalities store ownership information, and forest companies keep maps of forest growth and cutting licenses. These data sets are expensive to collect and should last for a long time. They are useful not only for the developer of the data sets but could be used by many other data users. To be usable, however, people have to be able to determine what information is in a file, they have to be able to determine whether specific data are available, where they are available, and if they are acceptable for a particular purpose.

Simply described, *metadata* are data about data. They describe characteristics of the actual data. Since data about data should make a small portion of the database, this information should be kept as high up in the hierarchy as possible. Here are some conditions for metadata:

- Data without meta-information are useless, at least outside the project for which they were collected initially.
- Metadata should be linked inextricably to the data that they describe, at storage, maintenance, and import/export.

This demands the development of standards for metadata. Metadata have to be presentable by the user software in connection with its host. When searching for spatial data and querying spatial objects (e.g., by pointing with a cursor), a parameter for the quality of the answer should be available. As far as possible, metadata should be used automatically, or at least optionally, in data manipulations, as for error propagation in the integration of two themes. Visualization of data quality should be supported by special cartographic routines.

Basic Cataloging   Metadata are often called the dictionary component of GIS, sometimes even the table of contents or the index for spatial databases. There are usually four access types that we can distinguish:

- *Spatial access* determines the availability and quality of information for any point of area.
- *Temporal access* determines the availability and quality of data with respect to time periods.
- *Attribute access* determines the attributes in a hierarchical manner, often in the form of a thesaurus.
- *Spatial resolution* determines the scale.

To decide whether a particular data set is suitable for a project, the lineage and the specifications of the data set are important, user costs and rights have to be determined, addresses for information and purchase have to be available, transfer options (formats, data media) have to be known, and the completeness has to be guaranteed.

Metadata and Meta Structures   The most general approach is developed in the *Contents Standards for Digital Geospatial Metadata* of the Federal Geographic Data Committee (FGDC). This document lays out the definition of metadata in detail. Ten categories are used:

- *Identifier:* description of the goal, the theme, spatial position, status, graphics, etc.
- *Data quality:* origin, positional and attribute accuracy, logical consistency, and completeness
- *Spatial organization of the data:* specification of the data model
- *Georeferencing:* projection, datum, coordinate system
- *Entities and attributes:* definition of entities and attribute keys, measurement scales and ranges
- *Acquisition:* sources, ordering procedures, specifications for digital transfer, transfer media
- *Contacts:* contacts, security classification, etc.
- *References:* suggested reference form: source, developer of data, title, owner, date of development and editing, links to other references
- *Time span:* how long the data are valid
- *Contact address:* people, addresses, telephone numbers, e-mail, working hours

With these categories, we can describe spatial data sufficiently. However, this standard only specifies the content of metadata and does not go into their codification and storage. To go further, there is the *Data Quality Report of the SDTS* (U.S. Spatial Data Transfer Standard). This states the five dimensions of data quality:

- Origin of data
- Positional accuracy
- Attribute accuracy
- Logical consistency
- Completeness

These statements can be given for all levels (database, theme, object, element/ primitive). The description can be undertaken in text form, by a list of attributes, or as a map that gives spatial differentiations in accuracy.

Object-Oriented Data Models

Object orientation is spreading into all aspects of information processing and brings advantages into many of them, besides being one of the buzzwords in computer-related discussions. Because of this multitude of effects, we have to be especially careful with the term. We have used the term *object* in GIS for years. Object is the computer implementation of the entity in the real world or the feature on a map. Some GIS researchers have claimed that GIS has therefore been object-oriented well before the term was invented. Such statements overlook that object orientation incorporates much more than the term *object* itself and includes structures, functions, and hierarchies. Thus, the term *object* has obtained a specification of meaning.

Object Identification An object is a software package that contains a collection of related procedures and data. In the object-oriented (OO) approach, these procedures are called *methods,* functions for visualization, transformation, and analysis. The data elements are called *variables.* Objects are building blocks of systems. They can be abstractions of spatial entities, but they can also be functions, for example. Each object contains its own description (coordinates and attributes) but also methods. The principal characteristics of the object-oriented approach are as follows:

- Encapsulation
- Inheritance
- Abstract data types

Encapsulation The combination of methods and variables in one unit provides a way to distinguish between internal and external aspects of objects where the internal aspects are hidden and are accessed only by the object's methods. Objects don't touch each other's data structures; rather, they send each other messages that call methods into action. These methods, in turn, access the required variables.

Methods Methods are thus procedures within an object that are made available to request and supply services. All communication between objects takes place through methods.

Messages Messages are signals from one object to another, requesting the receiver to carry out one of its methods. A message consists of three parts: the name of the receiver, the method that is to be carried out, and parameters that the method requires to fulfill the task.

Class and Inheritance Many objects are similar to other objects. Highways, roads, and trails have many things in common and differ in only a few characteristics. Therefore, there would be much repetition if it weren't for classes. Objects are grouped into classes. Classes are templates that define the methods and variables that the objects in one class have in common. To look at the class–object relationship, an object is an instance of a particular class. Its methods and variables are defined in the class; its values are defined in the instance, the object. Thus, the object inherits the methods and variables from the class.

Generalization   Classes can be grouped into superclasses (generalization) and divided into subclasses. Each class contains the properties of the superclass (inheritance), with some other properties specific to the particular class.

Abstract Data Types   Early programming languages had a fixed set of built-in data types. Even though these data types were fairly extensive, they were all defined by the way information is stored in the computer and had little relationship to real-world objects. Modern programming languages let the user define new data types, called *abstract data types*, by combining existing data types in new ways. This data abstraction has limitations, however, because programmer-defined data types are not treated in the same way as built-in data types are. Object-oriented languages are designed to be extended and adapted to specialized needs through the development of abstract data types. The tool for creating new data types is the class.

Time in GIS

Even though time is a very important factor in the handling of geographic information, it has been given very little attention until relatively recently. Sinton's (1978) framework of geographic data representation describes location, time, and attribute as the three components of geographic information. Imagine a three-dimensional space consisting of location, time, and attribute. The data that we use always have one of them as constant, the second varying in a controlled way, and the third measured (Table 4.3). As can be seen, tables and sequences have the location fixed, but maps

Table 4.3   Framework of geographic data representation

Information Type	Component Type		
	Fixed	Controlled	Measured
Soils data	Time	Attribute	Location
Topographic map	Time	Attribute	Attribute
Census data	Time	Location	Attribute
Raster imagery	Time	Location	Attribute
Weather station reports	Location	Time	Attribute
Flood table	Location	Time	Location
Tide tables	Location	Attribute	Time
Airline schedules	Location	Attribute	Time
Moving objects	Attribute	Location	Time

fix time. This is not a generic situation but has developed over the decades due to neglect of the time dimension in spatial research. In the following, the major GIS functions are studied with respect to an expansion into the temporal domain.

Inventory   The purpose of an inventory is the enumeration and description of entities that are considered important. The enumeration takes place at a point in time and in a given area. This is very much the approach that is taken for most databases. What is a record in a standard commercial database is a region (polygon) in an inventory. A file is the entire area at a particular point in time—the *snapshot model* in the spatiotemporal literature.

Is it possible to expand on the time horizon in this scheme? Can the shift be shown more dynamically rather than in such a stringent, discrete manner? There are ways. Each region has a starting date (a date of birth) and an ending date (a date of death). When an area changes its boundaries, the old polygon is archived and one or more new ones set in its place.

Analysis   Geography has been interested in dynamic processes for most of its existence. Erosion cycles and sequent occupancy have been early examples, models of diffusion, ecological spread functions, and spatiotemporal autocorrelation are examples of modern spatial sciences. For some time now GIS research has studied change through comparison and subtraction. These studies are based largely on polygon overlay techniques. The data volume is the critical factor for these approaches. Research in spatiotemporal systems would classify all these approaches as snapshot models. More sophisticated approaches are still awaiting discovery.

Updates   The substitution of current material for outdated material is a typical temporal procedure. Many practical processes have been developed for the aspatial situation, and incremental updating procedures are available for GIS maintenance. The best known might be the weekly reports called "Notice to Mariners" and "Notice to Airmen," which update the nautical and aeronautical charts. This has found an equivalent in the spatiotemporal literature through the *update model,* which stores only those components of a database that have been deleted or added in the reporting period.

Quality Control   Temporal systems increase the complexity of quality control significantly.

Imaginary Error   If a comparison of two data sets at different times shows discrepancies, they might not be errors but the unexpected results of some temporal processes. To distinguish such situations from real errors, quality control has to incorporate an (automated) understanding of the process.

Lineage   Errors can be time dependent (i.e., what is an error today might not have been one yesterday). Data therefore have to have time tags with respect to their collection, correction, and manipulation.

Scheduling   Databases can benefit greatly from scheduling capabilities. As certain conditions change, these changes can trigger other conditions. A simple example is the triggering of coverages in themes when change of scale occurs. More complex ones could be electronic charts that change scale as ships come closer to shore, or analyses of map revisions that determine the date for the next revision (i.e., the more changes in the present revision, the sooner the next revision has to be scheduled). Since change is inherently temporal, so are triggers.

Display   It is not easy to display temporal processes. This has little to do with the lack of research in the problem but much to do with the inherent problems of mapping space and time in a two-dimensional medium. At this time, four mapping techniques are available for the description of spatiotemporal processes:

- Time sequences (e.g., multiple editions or time series)
- Change data (e.g., text, graphic, or digital additions to a base representation; the "Notice to Mariners" and "Notice to Airmen" are examples of text)
- Static maps with thematic symbols of a temporal theme (e.g., dates, rates, growth)
- Animation, where spatial information is shown by many maps that are sequenced through time

GIS Analysis and Operations

It might be useful to develop an overview of the functionality of GIS or to establish a taxonomy of GIS functions. One of the most logical views of GIS functions follows the dictum—originally from computing science—that every procedure transforms data from one state to another. If we do this for the basic spatial features, we get the results shown in Table 4.4.

It is clear that this does not encompass all the GIS functions. The frequent occurrence of generalization (one of the areas that GIS has largely left out of the mainstream of development) indicates that some of these cells might better be left empty. There are other classifications that are based on the functionality of GIS and/or some differences in the treatment of data in databases. Others start again from the "tasks" to the functions that are needed to fulfill the tasks.

Map Algebra

There is one view of spatial processes that has had more success and has survived a sufficient number of years that it can no longer be considered a fashion. It is identified with several names, map algebra, cartographic algebra, cartographic modeling, and so on. The terms suggest that there is a formal language for spatial processes, similar to mathematical–algebraic language. The elements of such a language would refer to cartographic models of spatial reality. The basic processes are transformations that are formulated as combinations of operators. Every operation transforms data from one structure into another; a transformation consists of one or more operators.

Table 4.4 GIS functionality

	From				
To	Points	Lines	Areas	Networks	Surfaces
Points	Generalization	Line generation	Thiessen, allocation, etc.	Node definition	Interpolation
Lines	Line simplification	Generalization	Polygon topology	Edge definition	Interpolation
Areas	Centroid, etc.	Circumference	Generalization	Generalization	Interpolation
Networks	Node identification	Edge identification	Allocation	Generalization	Interpolation
Surfaces	Raster, TIN	Contours	Zonation	Channel and ridge networks	Generalization

As with higher-order programming languages and general algebraic notations, map algebra tends to abstract from the detail of data storage and the implementation of individual operators. Cartographic algebra should therefore be independent of the basic data structures (vector or raster). The first definition of a "language" for map algebra came from Tomlin (1990), initially also at "the Lab." A basic idea of map algebra, as in mathematical algebra, is the use of variables; spatial themes or layers are given names and are used as variables in expressions. Operators therefore act on one or more themes, the result being another spatial theme, and in most cases all themes are congruent (i.e., have the same cell size and matrix dimensions). The process that combines operators with themes is called *operation.*

Categories of Operators   Operators in systems such as GISs have similarities that can be grouped. Cartographic algebra is being categorized on the basis of the spatial processes that the operators perform. *Local operators* create a new object from one or several objects at the same location. For example, a raster pixel is defined as "suitable for a rural airport" if the same pixel is flat in the slope coverage, is determined "solid" in the soil coverage, and so on. Most polygon overlay procedures are local operations. *Neighborhood operators,* called *focal operators* by Tomlin, create an object from objects at the same position and the immediate neighborhood. Most computations of slope fall under this category.

Regional operators, also called *zonal operators,* use regional boundaries to delimit the area that contributes to the creation of an object. Statistical analysis usually works with statistical areas as the basis. As an expansion of zonal operators, Tomlin mentions *incremental operators:* for example, slope or exposure. *Global operators* create objects using very large areas, often the entire region given by the spatial database. Visibility analysis and contouring belong to this group. As we move from local to global operators, the computational complexity of the processes increases. Local operations are directly proportional to the number of objects. So are neighborhood operations, even though the overall process has to be multiplied by the number of neighbors. But global operations usually increase at the square of the increase in the number of objects.

Spatial Filtering

Filtering is conceptually the simplest way of transforming data from one state to another. All that is needed is a transformation formula for every element. The important aspect of filtering is that the formula is exactly the same for every component of a database. This usually implies that the elements are relatively simple in themselves. In GISs, filters are usually applied to raster data.

The simplest method of spatial filtering a continuous surface is passing a square window over the surface and computing a new value for the central cell of the window by adding up all weighted cells of the window. The function operates on one pixel at a time, using the pixel itself and some surrounding pixels. In the example in Figure 4.15, taking a pixel and its eight neighbors, adding them up, and dividing them

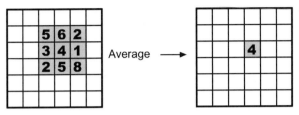

Figure 4.15 Spatial filtering.

by 9 will give us a raster that appears much smoother than the original raster. Since we have to perform this operation one pixel after another, this process is usually called a *moving window*. Filtering is the classical example of neighborhood operations. The neighborhood is fixed: The pixels that contribute to the value of the computed pixel are always in the same position relative to the pixel in question (Figure 4.16).

To present a simple classification, we distinguish between two basic classes of filters: filters that smooth the image and filters that accentuate, contrast, and enhance breaks. The first suppress local detail, treating it as noise, and focus on the large-scale variations. Averaging the window would be one example. The second works just the other way: Local detail is emphasized and large-scale variations are suppressed. Smoothing filters are also called *low-pass filters* because the low frequencies of the image surface are maintained. *High-pass filters,* on the other hand, emphasize the high frequencies of the image surface.

Here are a few general properties of filtering: Filtering is irreversible (i.e., inverse filters usually don't exist), and filters can be applied to the data set multiple times. Usually, that means that the effect is fortified; filters are often applied before or after other analytical processes (i.e., as preparation or correction). For example, after a classification in remote sensing, one often applies a low-pass filter to suppress singular pixels.

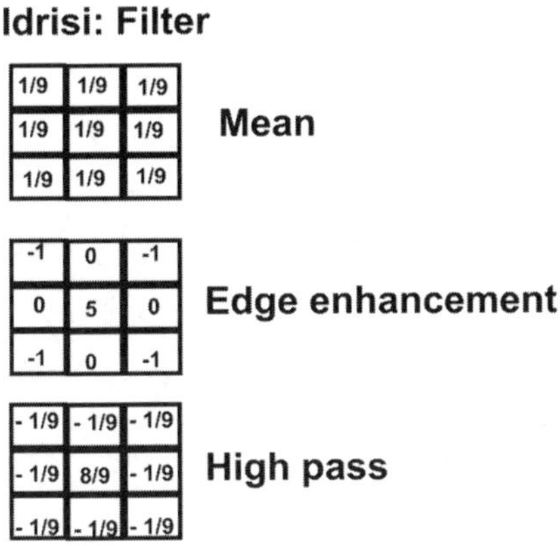

Idrisi: Filter

1/9	1/9	1/9
1/9	1/9	1/9
1/9	1/9	1/9

Mean

-1	0	-1
0	5	0
-1	0	-1

Edge enhancement

- 1/9	- 1/9	- 1/9
- 1/9	8/9	- 1/9
- 1/9	- 1/9	- 1/9

High pass

Figure 4.16 Filtering.

Interpolation

Interpolation is the procedure of predicting unknown values using known values at neighboring locations. When collecting spatial data, we are usually forced—for economic as well as fundamental reasons—to sample the area (i.e., take measurements at distinct places, usually points). However, it is often expected that we deduct from the measurements some information about the remainder of the study area. This deduction can use different approaches, depending on the assumption that we have made about the surface. As it has been mentioned before, we basically have to deal with two assumptions about the continuity of the data distribution of a layer:

- *Tessellation assumption.* The points between sampling units are assumed to take on the values of the closest sampling unit, sometimes within some boundaries. In other words, space is divided into regions.
- *Field assumption.* The points between sampling units are assumed to take on values that are weighted averages of the surrounding values with a multitude of weighting functions to choose from. This will result in smooth surfaces.

The first case is usually given by the set boundaries or satisfied by triangulation, the second by interpolation (Figure 4.17).

In the context of GIS, interpolation is usually considered a two-dimensional problem, where a surface is developed from a set of points—rarely, lines. Let us call the points that are known (with their $x,y,$ and z values) *data points*. Ideally, data points should be distributed as evenly as possible throughout the area that is to be interpolated to preserve the quality of the interpolation. Since we cannot create a value for every place on the surface (that would amount to an infinite number of values), we will arrange the interpolation according to the data model of the DTM in question. This is usually a raster, even though other models (e.g., contours, piecewise continuous surfaces, even sets of triangles) are possible. This has two outcomes. First, the number of points interpolated is usually much larger than the number of data points.

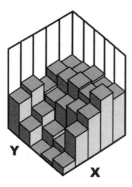
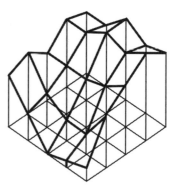

Figure 4.17   Interpolation.

If the data points were well chosen (i.e., along breaks on the surface), this makes a lot of sense. The other point is that we usually have to go two steps to get from the initial point of selection to the graphic representation: first, interpolation to a raster, and then, contours, perspective views, and so on. This is also the case with triangulation.

In the proper use of the word, interpolation is restricted to the estimation of grid points situated inside the set of data points usually called the *convex hull.* The estimation of points outside that set is called *extrapolation.* Estimates outside the data set are less reliable and prone to higher error rates. Like interpolation, extrapolation is guided by assumptions that are made about the surface, but since these assumptions are usually specifically for the area within the data points, some unexpected results can happen outside.

So far, we have been talking about interpolation of surfaces of ratio and interval data. There is also the question of interpolating nominal samples. How do we create a map from a set of soil samples, for example? We deal with this issue in more detail when discussing triangulation.

Global Interpolation One major differentiation among interpolation procedures is the number of data points used for the computation of one value. Most procedures use a small number (e.g., six) of immediate neighbors, but there are some that use all data points available. These *global interpolation techniques* give us dominant general trends and ignore local behavior. We discuss one that is based on polynomial functions, usually called *trend.* This is the one used most frequently.

Trend surfaces are polynomial regression surfaces (Figure 4.18). Each function

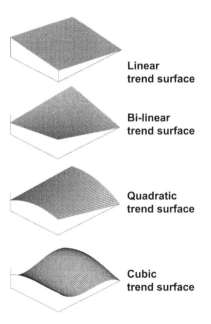

**Linear
trend surface**

**Bi-linear
trend surface**

**Quadratic
trend surface**

**Cubic
trend surface**

Figure 4.18 Comparison of surfaces.

represents a dependent variable, usually height (z), and two independent variables, usually the horizontal dimensions (x and y). The simplest case is the linear trend surface, whose equation contains the three parameters a, b, and c that have to be computed. The method used to compute these parameters is the regression method.

$$z = a + bx + cy$$

Such an approximation results in a plane that does not go through the data points. From an empirical standpoint, one usually argues that the data points carry a fair number of errors, and given the assumption that reality is smoother than the data themselves, the best guess is such a surface. An alternative explanation is that the data comprise a multitude of variables, and since we cannot evaluate all of them, we are only looking for the trend that the data suggest. If the linear case is not plausible and we can assume that our real surface has curvature, we have to change to a use of non-linear functions. We will not go into this further.

Local Interpolation   Local interpolation computes values for grid points by using the data points in the immediate neighborhood of each grid point. Neighborhoods are rarely clearly defined; They usually depend on the density of points. Most local interpolation procedures use the six closest data points. Well, not quite: Since finding the six closest points is more expensive than searching in a given area around the grid point, one often searches in a given region and changes the size of the region only if the number of points gathered does not fall within a range around 6 (e.g., $4 < n < 8$). As a rule, we can say that the larger the number of neighbors that are used, the smoother the surface becomes. If all data points are used for the computation of all grid points, the surface will be smoothest. However, the resulting surface will still look different—and usually less smooth—than the trend surfaces through the same points.

Exact versus Approximate Interpolation   An important criterion that distinguishes between different interpolation algorithms is whether the interpolated surface exactly matches the height of each data point or only passes nearby. The difference is based on different expectations about errors in the values of the data points. Exact interpolation ignores errors and creates a surface that incorporates them. Approximate interpolation algorithms put more weight on the smoothness of the surface and therefore come only as close to the data points as the basic conditions allow.

Inverse Distance Weighting   Inverse data weighting (IDW) is the most frequently used interpolation approach. It is based on the assumption that surrounding points should contribute to the value of the grid point, with the closer points contributing more than those farther away (Figure 4.19). After a number of data points have been selected, the distance is computed between the grid point and the data points. Then the value of the data point is multiplied with the inverse of the distance;

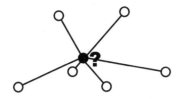

Figure 4.19   Inverse distance weighting.

these values are added up and divided by the sum of the inverse distances. In many cases the square or cube of the inverse is used, to guarantee continuity.

Spline Functions   Before the computer era, some very basic tools were used by drafters to draw smooth curves. Tags were pushed into the drafting board along a curve, and a thin wooden strip was laid into the tags. This strip was called a *spline,* and this is the name of the digital process that imitates the spline. Such a wooden strip would attempt to minimize the tension on the material by minimizing the curvature of the line. The design of ships was one application of spline curves. Spline curves do not set a maximum curvature but a maximum rate of change of the curvature (a maximum tension). There is an entire family of spline curves that can be used, depending on the tension demands. Spline curves are usually fitted together from pieces using overlapping sets of points.

Recently, a number of free-form curves have been developed which give the user more freedom than splines do. These are called *Bézier curves, B-splines,* and *NURBS* (nonuniform rational B-splines). However, they have not conquered GIS programs because the word in GIS is that lines should go through data points, something these curves do not guarantee all the time (except for Bézier curves).

Optimal Interpolation: Kriging   Kriging is performed in two stages: first, computation of the basic surface behavior, and then, interpolation with one of the Kriging techniques, using the appropriate *semivariogram,* which expresses the surface behavior. The process starts with computation of the semivariogram. This is done from a sample of data points and gives basic information about the change in the variance with increasing distance. For a specific distance (a lag) h, the variogram is the average squared difference between pairs of values (z) at the lag h. From these empirical functions, the model is selected that fits the curve best. There are many such models; most important are decisions as to whether one should consider local effects (*nugget effects*) and systematic trends (*drifts*). This needs some knowledge of the distribution of the data points and the type of surface.

One of the most valuable results of a kriging interpolation is the fact that we not only get a map of the interpolated surface but also a map of the errors of estimation, which are highly dependent on the distribution of data points. This is not such an important issue for tasks with many points (topography, etc.), but for situations where every sample point counts (e.g., drill holes), this can be a very good guide to where one might need more information.

4.4 DISTANCE-BASED OPERATIONS

Waldo Tobler, by many considered the "dean of GIS," formulated what he called the "first law of geography" (Tobler, 1970): "Everything is related to everything else, but near things are more related than distant things." Tobler applied this law to human behavior when he phrased it, but it has value in all other areas of spatial analysis.

Buffers

The influence of an object on its surrounding has been an issue of interest in geography for a very long time. Based on studies of human and animal behavior, *action fields* were developed. Action fields are based on the frequency of contacts on the basis of distance. Following Tobler's law, the average density of communication decreases with distance. This is true for trips away from home, shopping behavior, car travel, telephone calls, and so on. The search for food among animals follows the same pattern.

Let us take the example of a road that has to be cut through a forest. The forest has considerable wildlife that should be considered in the planning stage. What influence will the road have on the activities of, say, bears? Every bear whose action field intersects with the road will be affected, but the annoyance will increase the closer the road is located to the bear's den (Figure 4.20). So, by studying the curve we can see what impact the road will have, and using different distances around the road, we can show how much of the wildlife will be affected. This is what we call a *buffer.*

Buffers have developed a life of their own. Few remember the origin of the technique, using a metrically simple figure. There are many official operations that use buffers:

- *Zoning laws.* Residential development has to be *x* feet away from industrial buildings.
- *Protection.* No logging may occur within *y* feet of water flows.

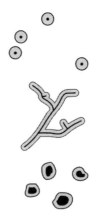

Figure 4.20   Buffers.

- *General selection before detailed study.* Include all trees within z feet of a road (potentially affected); include all people living k kilometers from a nuclear plant (e.g., higher rates of leukemia).

Buffers are constructed directly within vector systems. Buffers around points are circles; around lines they are corridors (with circular ends); and around areas, they become zones.

4.5 OVERLAY

Overlay is one of the central themes in GIS. Some have argued that overlay capabilities make the difference between a GIS and a graphic/CAD system. That is arguably true for the early years of GIS. *Overlay* means bringing together regional information on multiple levels on the basis of coordinating functions. Overlay is one of the areas where the difference in the spatial data types used (vector or raster) makes a great difference. It has therefore also been the field of overlays where many of the philosophical battles in GIS have been fought.

Multithematic Analysis

From its beginning, GIS programs have had data organized in *layers* (the original term), also called *variables* (in mathematics and traditional geography), *coverages* (ever since Arc/Info became available), and *themes* (ArcView). As mentioned before, this agrees strongly with the habit by most disciplines to dissect reality into components and specializations. This is the simplest way of recording our environment; object-oriented approaches demand much more understanding from the data recorder.

The central topic in overlay analysis is that of selection—often combined with statistical aggregation—of objects in a theme with selected or interactively generated objects that are physically or at least logically of a different theme. The first selection, the often used *point-in-polygon* (PiP) *search,* asks for the object that surrounds the point. The others are areal selections. We can ask two questions here: for all objects that are completely inside the search polygon and for all objects that have any portion within the search polygon.

Raster Overlay

In 1971, Ian McHarg published *Design with Nature,* a book that had a profound impact on the professions of landscape architecture and later GIS. The book presents overlay as a method to solve planning problems. Not that he invented the method—it had been practiced for some decades before—and not that he suggested having it done by computer. But the timing was perfect for the computerization; it just needed a good description with some convincing graphics to do it.

The first overlay programs were raster systems. It is considered the simpler approach: One pixel each time in one layer is compared with pixels at the same location in other layers. We have to be aware that there is a very practical limit to the accuracy of raster maps. Nothing smaller and/or more precise than the cell size can be shown. This is a blessing in disguise because many of the errors that the vector approach has to battle with will disappear.

Vector Overlay

Conceptually, the vector approach is very similar to the raster approach. What is added is the intersection of the polygons in the various layers and the creation of an attribute table for the new polygons. However, the fact that vector coordinates are infinitely detailed makes the smallest error a problem. If a line is digitized twice in two unrelated processes, rarely do the two lines match perfectly. Usually, they meander on top of each other, giving what we call in the profession *sliver lines* or *spurious polygons*. Michael Goodchild formulated what we often call the *Goodchild paradox:* The more accurate are two independent digitizations of the same line, the larger the number of spurious polygons. The size of the spurious polygons becomes smaller— both the individual polygons and their sum—but there are more and more of them as the accuracy increases (Goodchild, 1997).

4.6 GENERALIZATION

In traditional cartography, generalization is necessary when a map at a particular scale is derived from maps of larger scales. Since features cannot simply be made smaller, such changes of scale involve more than the reduction of size. Very often, shapes have to be simplified, wiggles in lines straightened, small shapes collapsed into fewer shapes, and many features deleted so that only those remain that are large

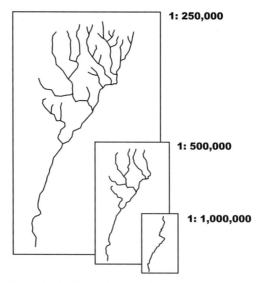

Figure 4.21 Generalization of a hydrological network.

enough or do not interfere with other features in relatively featureless areas (Figure 4.21). Generalization in traditional cartography usually means that all data are given in one scale and have to be transferred into another.

One aspect of this is what we call competition for space among map features. Since representations have to maintain a minimum size to be readable, the relative size of features increases in relation to the space they cover in the real world. This eventually results in the shifting or even the outright deletion of features. The term *scale* is used in its broadest sense. Scale is not only the ratio of a measurement unit on the map to the same extent in reality, it can relate to the detail that is to be replaced. A wall map in a school is obviously less detailed than a topographic map of the same scale. Such comparisons make it difficult to work with the term *scale* alone.

In the digital arena, generalization originally meant reducing the volume of data to save storage space and processing time. Not much else was considered possible. Selection was done manually, simplification was largely ignored. Generalization is still a young field in GIS and a large body of development is needed to come near an automated generalization.

Network Analysis

Networks serve for the display and analysis of connectivity and sequential processes. Therefore, they not only serve to demonstrate spatial phenomena such as street networks, power grids, or river systems, but also abstract relationships such as work schedules and personal hierarchies.

For geographic information systems, two aspects of network analysis are especially important: the connectivity of spatial phenomena and movement in space. Movement can occur in two ways: in a predominately areal or a predominantly linear fashion. Predominantly *areal movements* such as the dispersion of contaminants are usually better represented in raster mode. The distances used are physical (straight-line) distances. *Linear movements,* as along power grids, river systems, and street networks, are well represented by vector-based nets. The measurement of distance takes place along the edges of the net and the units can be miles and kilometers but also time or cost, for example. As-the-crow-flies distances are used for comparison, nothing more.

In a third group, a *combination of areal and linear movements* occur. For example, when studying the flow of rainwater, we begin with an areal collection of water which then congregates into creeks and rivers. Commuter travel could be seen similarly; walking to public transportation can be seen as an areal activity (because it takes place in a dense maze of streets and walkways) with straight-line distances, but the transportation patterns are clearly linear.

4.7 GIS AND MODELS

For the GIS community, models are structures that apply some knowledge to a particular spatial problem and use GIS for the solution. The emphasis is on a predefined structure that guides the analysis. It usually starts out with methodical models or theories, but most of these theories are quickly applied by thematically oriented disciplines. We have seen this with the Thuenen theory, which Thuenen himself applied to agricultural production in the nineteenth century and others applied later to urban growth. In the following we discuss some fields that have used spatial models to a large extent. There are, of course, many more.

Ecological Models

Ecological models are models that represent the interrelated agents in a natural household. Such models are used, for example, for biotic production processes. The range goes from fish production and forest management via agricultural planning to irrigation processes. Even though the latter topics do not seem to fit into the spectrum of ecological applications, they do share the assumptions of an approach that is system oriented, natural household based, and process oriented.

The scale of environmental modeling depends on the type of ecological approach the researcher is taking. Population ecologists usually work at the local level, in micro scale, whereas landscape ecologists look at the regional level, the macro scale. Early, simpler models used simple layer structures and defined relationships between layers as elements. Later models, which are more process oriented, had their origin in aspatial approaches and developed into powerful tools for presentation of the exchange of energy and matter. Here are some of the main directions that ecological modeling has taken:

- *Vegetation systems* simulate the succession of vegetation groups over several states.
- *Habitat modeling* studies the potential distribution of animals in dependence of local and structural components of regions.
- *Nutrient cycle models* analyze the locational and growth conditions of the exchange of energy and matter.
- *Predator–prey models* study and simulate the relationship between predator and prey animals. These models show that there has to be a stable population in both groups for them to exist.

Ecological models are among the most complex and demanding systems with respect to their structural conception and their empirical base. An interesting example of an ecological model is the simulation of forest succession. Starting with the fact that Douglas firs can withstand forest fires very well whereas hemlocks are very sensitive to them, forest succession was simulated on the basis of whether or not forest fires were suppressed. The results were very interesting: When forest fires are suppressed, hemlocks will thrive and eventually overpower Douglas firs. The climax tree will be hemlock (climax vegetation is the ultimate vegetation, i.e., the vegetation that will dominate a mature forest). When forest fires are not suppressed, hemlocks will perish and Douglas firs will grow and become the dominant trees.

An interesting lesson can be learned from this simulation: Fighting forest fires creates forests that are dominated by hemlock. However, hemlock has always been considered the commercially "undesirable" species, whereas Douglas fir is the precious tree. In other words, fighting forest fires, which was introduced when forests were discovered to be of commercial value, works absolutely against the interests of commercial use.

Atmospheric Models

Atmospheric models are probably the best developed and most frequently used models. Meteorology and physics are the main disciplines that use these models, but other fields use them as well. Distinctive characteristics for the various models are their extension and resolution or scale:

- *General circulation models* (GCMs) are small scale and predict the atmosphere in the near future (short- and medium-term weather prediction, meteorology) or predict the atmosphere in the long term (climatology).
- *Climatic models* at a regional level are usually based on the interpolation of empirical series which are registered at local stations.
- *Boundary layer models* focus on the interaction of the atmosphere with Earth's surface, emphasizing phenomena of the biotope. The atmosphere is seen as the carrier for thermal flows and moisture which controls the basic physical, chemical, and biological functions on Earth's surface.
- *Atmospheric diffusion models* also have at the center of attention the atmosphere as the carrier of dangerous substances. The sources can be point, line, and area sources.

Hydrological Models

Spatial models in hydrography have a long tradition that reaches back beyond the development of GIS. Before GIS could provide sufficient general structures for the processing of spatial parameters, hydrological models were managing terrain, precipitation, soil, and other relevant data in very specific structures and formats. Also, the modeling and simulation techniques for dynamic models were developed within hydrology without any input from, or consideration of, other disciplines, especially geography and GIS.

The application of hydrological models falls into a number of relatively specific areas:

- Surface flow models start with precipitation and consider the interaction of precipitation with the various layers of vegetation, the percolation of water, and the surface runoff of the remainder. Each of these quantities is treated in a storage with limited capacities and transportation of the overflow to other storages. The stepwise simulation has to pay attention to the spatial discretization (since most of these models are based on raster systems). This way, a limited three-dimensional water cycle can be simulated.

- Erosion models study the effects of surface runoff since surface runoff is the process that controls erosion. They analyze the process and the statistics of runoff for the estimation of erosion potential.
- Flow models, which simulate water flowing in large areas in varying directions, deal with wave patterns that are induced by wind or tides. The process often uses cellular automata.

Groundwater models are among the most important hydrological models. They usually start from a stable situation with the gradients, quantity and speed of flow, and porosity in equilibrium. Questions that can be answered are where to allow water pumps and the quantity of water that can be taken out without lowering the groundwater level beyond a certain level.

There are two schools of thought in hydrological modeling, one that develops *lumped parameter models* with a strong statistical orientation (with spectral and time series analysis and diverse multivariate estimations) and the other which considers *distributed models* with spatially distributed multitudes of variables. The latter is gaining ground because of the spread of GIS and the better forecasts it provides due to a closer attention to the physical processes.

4.8   SPATIAL DECISION SUPPORT SYSTEMS

Decision support system (DSS) is a term that is handled very broadly (for many, too broadly). When transferred into the GIS sphere, it is called a *spatial decision support system* (SDSS), with at least the same unnerving breadth of meaning. Even though most applications of GIS do support decisions, such a wide definition has limited usefulness. We therefore limit it (somewhat arbitrarily) to a set of quantitative methods that create a new level of spatial understanding. SDSSs support the user's decisions by giving alternatives and weighing them.

DSSs are designed to solve problems where the objectives of the decision maker and the problem itself cannot be fully defined. Such systems enable the user to combine analytical models. They help the user to explore the available options by applying the models in the system to generate a series of alternatives. They support a variety of decision-making styles and are adaptable to provide new capabilities as the needs of the user evolve.

So much for DSS directly. To get to a SDSS, we have to add a few more items: SDSSs have to provide mechanisms for the input of spatial data. They allow representation of the complex spatial relations and structures that are common in spatial data. They include geographical analytical techniques (including statistics). They provide output in a variety of spatial forms, including maps. Some of the problems that fall within this description are the search for optimal locations of retail stores, the best layouts for building sites, optimal highway corridors, and the division of areas into multiple land uses.

4.9 VISUALIZATION: THE NEW CARTOGRAPHY

Next, we would like to show that GIS can change things in a fundamental way. We will take cartography as an example. Traditional cartography differs from GIS not simply in the use of different tools but also in a distinctly different philosophy.

Everybody uses maps. We look at road maps when we are in a car, topographic maps when we are hiking, and subway maps when we are in a large city. We use maps to determine which way we have to go, what we will find when we get there, and what other information we have to collect to get to know the place before we get there. This is called *way finding*. There are many people who enjoy collecting information about the next vacation almost as much as spending their time at the place chosen, and programs are available that help us collect the data and view them. The map becomes a repository of spatial information. Then there is the professional use of maps. Maps are developed for decision making. Different layers of spatial information are combined to one map, often after some sophisticated analysis of the data. Professional maps don't put that much emphasis on the aesthetics of the image but more on the analytical powers of the software.

The Threat

From an editorial in *Cartographic Journal,* Vol. 30, No. 2, December 1993, p. 89 (special issue on map design):

> Debate at the Society's 29th Annual Technical Symposium...revealed the unease with which cartographers greet the explosive growth in map production using GIS and desktop mapping packages. There could be, it was felt, a weakening of the role of the cartographer if recognition of the importance of cartographic principles, exemplified by the fundamentals of map design, were to be overlooked.

> John Keates reminds us that, although our contemporary digital revolution is affecting map design, we have, in fact, gone through more fundamental changes to graphic representation in the past, even in the last 30 years.

These two quotes show the general attitude of the cartographic community vis-à-vis GIS: It is a defensive one, trying to guard against the perceived threat and belittling its influence. There is indeed a threat. Probably 90 percent of all maps are being produced by noncartographers, and the mapping programs (GIS and dedicated programs) are getting better. Not that the knowledge of the cartographer is becoming obsolete: There are many areas where the lack of mapping knowledge can become obvious and where at least some cartographic training would be useful, if not necessary.

Change in cartography is an interesting subject to study. Cartographers have always been quick in accepting new technologies: different drafting techniques, registration methods, and finally, the computer as a new drafting and archiving tool. Consider the mapping of of volcanic events for Mount St. Helens in the state of Washington. When the mountain erupted, a large portion broke away and slid downward. To re-create that event cartographically, because tools did not exist that could portray the event effectively in an animation sense, several individual scenes had to be hand drawn.

To reproduce the sequence of events was a time consuming task. The task of calculating dimensions and the extent of the area affected was also difficult. Cartographers have taken these technologies solely as more efficient tools for the production of maps but have been slow to change the structure and the outlook of their discipline when new technologies suggested such changes.

Technology does not only replace one tool with another. It alters philosophies, policies, the way we live. When barter was replaced by money, we did not just have an easier way to calculate value—our entire economic system changed from very small exchange units to very large trade systems which eventually expanded into a global economy. When better clocks and watches allowed the secure determination of longitude, it not only reduced the number of shipwrecks but changed seafaring from a risky adventure by daredevils to a routine marine business. We want to argue here that the introduction of the computer into cartography has changed the discipline like nothing before. By changing the practice of cartography, we change the way we look at maps; even the term *map* needs a completely new definition. An example of this can be found in some of the earliest maps for the Tahsish-Kwois Study Area located at North Vancouver Island in British Columbia, Canada, where two-dimensional representation (or traditional cartographic representation) gives way to three-dimensional mapping to provide a more realistic sense of distance and space—thus how we look at the map (Figures 4.22–4.25).

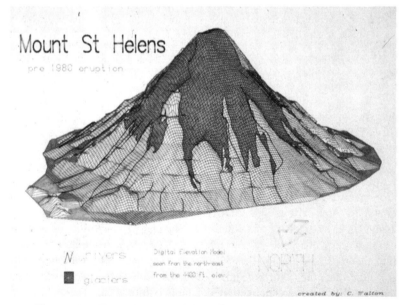

Figure 4.22   Thematic model of Mount Saint Helens (Pre-1980 eruption).

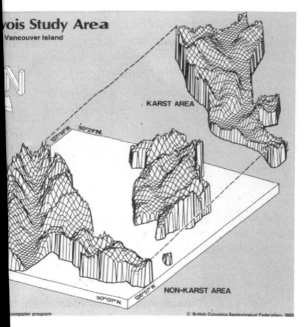

Figure 4.25 Tahsish—Kwois study area.

design is the declarative process in the preparatory stage
feature that will go on the map has to be known as to type,
before the first map element is placed. The map has to be
re the first line can be drawn, because any change of mind
ographer to start all over again. Alteration is the most diffi-

n today is undertaken predominantly on computers. Since
a graphic image of the data, a drafting program is prefer-
easier to add and delete different features and to change
mbolism. Computer-supported cartography works entirely
ography. Alteration is the easiest part. One starts with dis-
en with arbitrary or default symbolism. If the color of an
is changed; if the density of objects is too high in a por-
cts are deleted; and so on. Some of these processes may be
e deletion of objects), but that does not change the prin-
e changed.

design has become an iterative process of improvement
ortant than the artistic abilities of the cartographer and the
xperiment. Without any study of this issue so far, we can
be safe to say that most people are able to tell which one
given a choice of two similar versions. Such an assump-
bert Pirsig, for example, writes in *Zen and the Art of Mo-*

Tahsish-K

Norther

Cartography by Donovan Whistler
NOTE: This map produced using the ASPEX

F

Mount St H

N.

g

Figure 4.23

Mount

Digital Elevation

portion

portion

from the 4400 f

VOLUME

post eruption = 53

top portion, post

= 19.08 × 10⁷ c

Figure 4.24

Design

In traditional cartography
of map production. Every
quantity, and symbolism
known in all its detail bef
will usually force the cart
cult part.

Cartographic productio
the main goal is to create
able to a GIS because it i
their position and their sy
opposite to traditional car
playing all data on the scr
object doesn't fit, the colo
tion of the map, some obje
done by programs (e.g., t
ciple that everything can b

To use a computer term
in which rules are less imp
willingness and ability to
only speculate, but it mig
is the better map if they ar
tion is not so farfetched. R

torcycle Maintenance (1974) that even though it is very difficult to give a definition of the term *quality,* everybody seems to be able to distinguish between items of high quality and those of low quality.

Iterative map improvement is a fundamental change in the practice of cartographic design and affects all its components. Whereas traditional cartography has to spend much time on researching and teaching multiple design rules that have to cover every aspect of design, computer-assisted cartography needs only a few rules; everything else can be handled by trial and error.

Iterative map improvement can be based on a smaller set of rules, and the comparison of map versions is visual and therefore intuitive. It can be argued that computer-assisted cartography is simpler and can be taught in a shorter time span. Judging from the large number of noncartographers who are producing maps, there is a demand for a more compressed cartographic curriculum. However, the profession has not yet obliged by developing and offering short courses for mapmaking. This would not reduce the prestige of the profession; to the contrary, it could be argued that a discipline gains by making itself more accessible to the public. For example, making the computer easier to use has certainly helped to make computing science a very influential discipline.

So the recommendation to cartographers is to develop a subset of the cartographic curriculum that can be understood by many and to write books about it and actively teach it to as many people as possible. Such a simplified cartography would clearly not obliterate the discipline. Rather, it would make the other 90 percent aware of the discipline and give them some support to produce better maps. It might put cartographers in the driver's seat and give them the prestige that they deserve.

4.10 DATABASES

In digital cartography, the database, not the map, is the central depository of spatial information. Only the database can be added to, corrected, and so on. Any changes that are necessary for a traditional map mean that the map has to be redrawn. With the database, this is not the case. The separation of data storage from data display offers the cartographer a wide range of possibilities. First, data display comes together with data management. Cartographers therefore become one of the groups of experts who have special knowledge for the management of spatial databases as a domain. Equally, however, the other groups have special knowledge that makes them knowledgeable in the display of spatial data.

4.11 WHAT A MAP IS

As suggested above, cartographers are relatively conservative in the perception of their discipline. This includes the definition of the map. For a cartographer, a map is spatial information printed on (usually, white) paper. If the map is shown on a computer screen, it is a copy of the paper version, usually at a lower resolution.

Ever since the concept of hypertext was developed, linking information together has become a component of all types of programs, including GIS. It seems obvious that the map is better suited than most for such an environment. Maps are usually not the only information carrier that one queries when studying spatial issues, and linking additional information to places on maps is a natural. This also changes the role of the geographer in the collection and presentation of spatial information. Traditionally, it was assumed that spatial information was collected by geographers and other spatial scientists, so that all the cartographer had to do was to put it together on a map. With the widening of this role, the cartographer has the opportunity to become the custodian of all spatial information.

4.12 MULTIMEDIA

The modern map has depth. The ability to dig deeper into data allows us to add detail. It also has breadth, which is provided through the availability of databases. Zooming in and zooming out is possible today in ways and at speeds that were not possible before. An example of this includes terrain models such as the one shown in Figure 4.26. While it is impossible to reproduce in this book, using a computer with specialized software, one can move or "fly through" the landscape, in this case the valley of Indian Arm. Modern software also provides the means to display orientation relative to the landscape surface. Azimuth and altitude may also be shown with respect to viewpoint position.

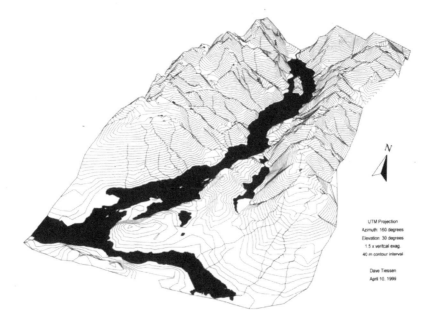

N

UTM Projection
Azimuth 160 degrees
Elevation 30 degrees
1.5 x vertical exag.
40 m contour interval

Dave Tiessen
April 10, 1999

Figure 4.26   Indian Arm, British Columbia, Canada.

Modern cartography can be identified by four components, two of which can be linked back to traditional mapping:

- Visualization
- Analysis
- Linkage
- Accuracy

Visualization

Visualization is different things to different people. It might be just another name for cartography; it might be the new, improved cartography, improved by technology and the promise of object orientation, open systems, interoperability, and globalization; it might also be the process of turning information into visual information. In our case it is the latter. Explaining facts by pictures and maps is an activity that has grown significantly in the last couple of decades. Still 20 years ago, modern geographers considered maps as an outdated mode; everything had to be done with quantitative techniques. Today, graphics and maps are very acceptable means to make information understood.

Analysis

Visual analysis might be the most potent type of analysis. Modern geography has neglected the power of visual analysis. However, viewing information in its spatial context is a tool of great potential, especially when combined with interactive mapping. Using computer mapping in association with some analysis is as old as computer mapping itself. SYMAP, for example, a widely used computer mapping program developed in the middle 1960s, was always thought of as an adjunct to quantitative analysis. The output was relatively unattractive and would not have been used for actual published maps were it not for the appeal of novelty. Nevertheless, at a time when one often had to wait months for maps from the mapping division, geographers and other spatial scientists flocked to the program because it gave an almost immediate rendering of the ideas of the researcher. Things have improved since then. Maps have become better looking and really publishable, analytical techniques have improved, and interactivity and animation are now possible.

Linkage

We deal with an almost infinite number of pieces of information in our digital world, and the important thing is not the existence of the individual pieces but the linkages between them. Only by linking units of information do they become useful; without linkage, they do not exist. We can say the same about the types of maps that are proposed here. The pieces of information are nodes that are combined to larger nodes, and as the size of the database increases, the nodes become less important and the links take over. Such a database is in constant change, nodes are renewed and replaced, and links are added and deleted.

The nodes are dependent on the links that relate them to other nodes. A node without a link is useless because it cannot be accessed. But too many links might not be helpful either. What we need is measures of information that describe not only the quantity but also the connectedness of the subjects. Equally, we should be able to identify areas in the networks that are better connected than others. Are measures of connectivity enough?

Accuracy

One of the arguments that cartographers use to protect their discipline from others, especially GIS, is that cartography is concerned about accuracy, whereas GIS is not. They claim that accuracy is one of the main pillars of the profession, whereas for GIS, accuracy is unimportant. There are many members of the spatial sciences that would vehemently disagree with this statement. Even though both cartography and GIS use data that come from different sources, are at different scales, and have been collected for different purposes, there is a great difference in the way they are combined. In cartography, data sets are usually combined in intuitively. When there are disparities, cartographic rules, professional experience, and common sense work together to create combinations that are logical. In GIS these direct interventions are not available, and much more has to be known about accuracy. This is why so much research has been undertaken in the areas of error propagation, scale and accuracy, metadata, and so on.

The surveyor is the guardian of accuracy. All hardware principles, control procedures, and geometric conversions have been developed in surveying or at least with substantial input from surveyors. Cartographers have not participated in these developments to a great extent. In fact, if accuracy plays any role in cartography, its absence seems to be more important than its presence. In cartographic generalization, the look of a map always comes first and accuracy second.

4.13 CONCLUSIONS

Let us conclude this chapter with a positive outlook. Contrary to the fear of many cartographers of being eaten up by other spatial scientists, we believe that cartographers can leap on the bandwagon of computerization as leaders and visionaries. Forget about accuracy; leave it to the surveyors, who are better prepared to understand it. Just use it the way it is offered and trust their judgment. Stop arguing over detail. Detail is necessary, but one does not have to talk about it.

Much of the defense of cartography these days consists of a litany of things that others should not do or be. Others should not draw maps; others develop spatial databases that are not good enough for cartographers. This is definition by negation, definition by restriction. Cartographers voluntarily declare themselves the accountants of the spatial sciences.

A more positive self-definition is easy to think of: Cartographers are the creators of visual information, the custodians of spatial information, the makers of good-looking maps. Cartographers design without the need for detail, cartographers construct without the need for accuracy. This, for all intents and purposes, is the position of the architects in the building sciences. Architects design, innovate, and create—without much care for detail, which is the engineer's problem.

One way to get to this position is to let the world know what cartography is—in the world's own words. Cartographers have to make their discipline understandable to all those who are creating maps today. To succeed, they have to simplify the field and teach this simplified version to the "masses." This will have several effects:

- There will be many more good maps in circulation, and poor design will have much less chance to succeed.
- Cartographic research will continue to be done by cartographers.
- Cartographers will become the undisputed leaders of all mapmaking, the architects of the spatial sciences.

EXERCISES

4.1. Name four types of ecological models and how GIS might be used in their application.

4.2. Both TIN and rasters may be used for modeling. What are the advantages and disadvantages of using each?

4.3. Discuss unstructured data and how information may be obtained for them. Provide two examples.

4.4. Under which circumstances does interpolation become a problem?

4.5. What are the differences between a TIN and raster data structure?

4.6. Discuss metadata structures and the elements that are considered to be part of these structures.

4.7. Assume that you have two raster grids of data for the same area. For one grid, each cell is 20 × 20 m. In the second grid each cell is 1 × 1 km. How would you go about filtering each grid? What are some of the considerations?

4.8. Name five uses for a digital terrain model.

4.9. We often see geographic information stored in grid format rather than in vector format. Why?

4.10. What is one advantage of using kriging over other forms of interpolation?

4.11. Why do philosophical battles about application of GIS tend to be oriented around the area of overlays?

4.12. What is meant by the term *generalization?*

5

GLOBAL POSITIONING
SYSTEMS

5.1 INTRODUCTION

Most of us remember a time not too long ago when we loaded maps into our pockets and vehicles for the purpose of navigating. Even when walking with a map, we were often faced with problems interpreting a map with reference to location. This was due in part to the type of map being used. Sometimes a flat projection used in the map did not represent the area adequately, or perhaps the map was outdated. The map could have been published many years ago, and many of the current features were not apparent. At other times cartographic liberties, including too much information, may have resulted in an overwhelming portrayal, making it difficult to understand. Features may have been generalized beyond recognition. Map interpretation is a subjective exercise and can lead to erroneous perceptions.

Consider an aerial photograph, charting your course as you move. The image must first be interpreted, which often is different from the site at the time you visited it. The aerial photograph may be a few years old or may have been taken at a time when foliage was at a different stage of development. Foliage or snow could be obstructing landscape features or new fences and buildings leading to disorientation. What may be quite apparent from an airplane may not be as readily apparent on the ground. Since maps and aerial photographs are two-dimensional presentations of three-dimensional landscapes, one is faced with navigating in a 3D environment using a 2D representation. Unless a pocket stereoscope is used and a stereo pair is present, there is no depth perception in a photo. Still, images in stereo from the air look very different from a view at ground level beneath a tree canopy or in a newly built up residential area. In some cases maps and aerial photographs can be viewed in tandem. Questions of accuracy must be considered. How accurate is the map? How accurate is the aerial photograph? Which is correct, which isn't? Many readers have experienced this.

Imagine that as you move you can locate your position as well as the azimuth between two locations and even how long it might take to move between the two locations. Add to that the possibility of making notes with specific reference to your travels. If this were possible, you would have a very useful tool. That is the global positioning system (GPS), or more specifically its value and usefulness. GPS receivers (sometimes called *GPS navigators*) can locate accurate positions anywhere on the planet. A navigator can provide coordinates (latitude/longitude or UTM), altitude, velocity, heading, and precise time of day in local or Greenwich mean time.

Most navigators also have a built-in mapping feature, sometimes called *moving maps,* that are displayed on large liquid-crystal displays. GPS navigators are capable of displaying positions relative to waypoints that were preprogrammed into them. They can plot routes, indicating where you have traveled or are moving to with respect to current position. Some models have street or waterway maps as backdrops and are also able to import other maps for various regions around the world. The backdrops serve as a means to orient the user as waypoints are recorded.

A GPS navigator will usually have a serial port for computer connections, allowing data collected to be downloaded, and some may have dual–serial ports. One port provides telecommunication and the other is used for downloading or uploading data. The ports may also serve as digital interfaces, providing a means to attach other sensors, lasers, and instrumentation or to connect a laptop with a direct link to running GIS or other software. Applications of GPS are increasing daily and include:

- *Climate:* air temperature, wind speed and direction, solar radiation, relative humidity, rainfall, snow depth, snow density, icing
- *Emergency situations:* temperatures, aerosols, contamination, radiation, vehicle speeds
- *Robotics:* movement, height, direction, speed, angle
- *Soils:* temperature, soil moisture, soil pH, soil organic matter, soil nutrients, electrochemical constant, porosity
- *Airplanes:* flight navigation, crop dusting, speed, direction, altitude
- *Human activities:* sight, sound, smell, touch, hearing, medical imaging, chemistry, sport
- *Water:* temperature, turbulence, biological oxygen demand, pH, level, depth, nutrients
- *Biology:* pathology, entomology, sunlight, leaf area, rooting
- *Construction:* movement, depth, quality, weight, distance, space
- *Noise:* indoor acoustics, outdoor, frequency, direction
- *Transportation:* speed, guidance, location, engine temperatures

5.2 GPS AND WIRELESS

Japan is currently the world's largest GPS car navigation system market, with 1997 shipments of 1.1 million units. Germany is another burgeoning car navigation system market, and the British Ordinance Survey is linking many operations in the field directly to GPS and integrating GIS in real time. Allied Business Intelligence, Inc. (ABI, Oyster Bay, New York), the world's leading supplier of business intelligence for manufacturers in the communications and emerging technology industries, predicts the worldwide commercial (nonmilitary) GPS receiver systems market will triple over the next six years, growing from an estimated $4.5 billion in 1999 to nearly $14 billion by 2005. Looking closer, the revenue for integrated circuits (ICs) used in GPS receivers was $148 million in 1999 and estimated to reach $2.2 billion by 2004.

Clearly, GPS will continue to be used in a growing number of spatial applications involving integrated geotechnology. It has been estimated that this will drive GPS communications applications to more than a 20 percent share for all geotechnology products by 2005. In the U.K., for the last half of 2001 it has been estimated that almost £22 billion have been spent toward enabling m-commerce in that country. Whereas e-commerce is dependent on hardware networks linked physically, m-commerce uses mobile connectivity. The assumption is that wireless networks and GPS will locate nearby stores and provide users with opportunities to shop on wireless phones (e.g., while driving I can locate which store has a radio I want to buy and how close it is).

The largest volume market for GPS will probably be its integration into the wireless handset environment. The FCC has mandated that all cellular phones must identify their location to within 125 m for 911 emergency calls. This Enhanced 911 (E911) mandate took effect in October 2001 and is driving the wireless carriers and handset manufacturers to find ways to meet the mandate's requirements. The market expectation is for hundreds of millions of GPS-enabled phones over the next five years. At the same time, G3 and G4 mobile products are expected to grow slowly. This slower growth has a number of causes, including confusion over the technology, cost, lack of content, and consumers seeking higher ergonomic integration into mobile products. The linking of GPS to telephones extends mainstream business applications, adding an entirely new dimension to consumer business closely tied to IT. GPS technology provides the best solution to meet or exceed the requirements of the E911 mandate.

Other estimates suggest that the GPS market in North America will grow 125 percent, to $4.6 billion by 2006 from $2 billion in 1999. These are astonishing numbers indicative of future potential growth based on GPS applications. They do not include associated and ancillary devices such as instrumentation and sensors, remotely sensed technologies coupled to GPS, linked visualization hardware and software, and the potential market for GIS-related technologies coupled to GPS. Although GIS will have continued and steady growth, it is GPS that would appear to be driving these spectacular levels of growth in the geospatial marketplace.

5.3 GPS HISTORY

The global positioning system (GPS) is an all-weather, continuously operating, satellite-based timing and ranging system. Officially known as the Navigation Satellite Timing and Ranging System (NAVSTAR), GPS is owned by the U.S. Department of Defense. GPS was initiated in the 1960s, and during the 1970s the original satellites were launched into space, forming the GPS block 1 system. Since that time other blocks have been added to the system (Joint Program Office, U.S. Air Force, n.d.):

- *Block I* (Rockwell)
    - 1974   Contract for eight block I satellites
    - 1978   Contract for three block I satellites
- *Block II/IIA* (Rockwell)
    - 1981   Contract for qualification satellite (GPS12)
    - 1983   Contract for 28 block II/IIA satellites
- *Block IIR* (Lockheed Martin)
    - 1989   Contract for 21 block IIR satellites
- *Block IIF* (Boeing–North American)
    - 1996   Contract and options for 30 block IIF satellites

In the early 1980s, block II satellites, considerably updated and improved, were launched and began operation. Initially there was no civilian use of the GPS system. Just prior to the launch of the first GPS satellites, the U.S. government had completed a new international geodetic system, WGS72. The World Geodetic System (WGS; initially WGS72) is the reference system being used by the GPS. Functioning for military purposes, GPS was used primarily to provide increased accuracy in navigation for military ships, aircraft, and weapon's platforms. NAVSTAR provides the following:

- 24-hour worldwide service
- Extremely accurate, three-dimensional location information (providing latitude, longitude, and altitude readings)
- Extremely accurate velocity information
- Precise timing services
- A worldwide common grid that is easily converted to any local grid
- Continuous real-time information
- Accessibility to an unlimited number of worldwide users
- Civilian user support at a slightly less accurate level

A similar Russian satellite system operated by the Russian Federation Ministry of Defense, called the Global Navigation Satellite System (GLONASS), has existed for a similar period of time. The primary missions of GLONASS are as follows:

- Air and naval traffic management, increasing safety
- Geodesy and cartography
- Ground transport monitoring
- Time-scale synchronization of the remote from every other object
- Ecological monitoring; search and rescue operation organization

Currently, the Russian Space Federation states that GLONASS will take into consideration the essential significance of the satellite navigation system for effective solutions in transportation, geodesy, and scientific and other applications, as well as the objective need for wide implementation of new information systems combined with satellite navigation users' equipment (GLONASS, 1999). On February 18, 1999, the president of the Russian Federation decided on Russia's involvement at a new level of international cooperation. That decree was meant to provide a national satellite navigation system as a basis for the development of international satellite navigation systems (GLONASS, 1999). However, a search of recent GLONASS satellite availability revealed that many of the satellites are nonfunctioning and that the system continues to deteriorate, with few operating satellites available for use (Table 5.1).

The current NAVSTAR/GLONASS operational status, indicating numbers and availability of GPS satellites, can be viewed on the Internet. There have been discussions in Europe surrounding the creation of a new, third satellite-based navigation system (EC, 2001). The Galileo GPS has now been fully approved. The Australian Cooperative Research Center in Satellite Systems (CRC) proposes FedSat, an Australian scientific microsatellite mission, a 58-cm cube weighing approximately 50 kg. It was launched in late 2002 from Japan by Japan's National Space Development Agency. Its purposes are as follows:

- To establish Australian capability in microsatellite technologies
- To develop the expertise necessary for sustaining those industries and profiting from them
- To test and develop Australian-developed intellectual property
- To provide a research platform for Australian space science, communication, and GPS studies

The U.S. NAVSTAR GPS system was not declared fully functional until mid-July 1995. The system was delayed by the *Challenger* spacecraft accident in the mid-1980s. Later, Delta rockets were used to place GPS satellites in orbit, and this has continued

Table 5.1   GLONASS constellation status, December 5, 2002[a]

GLONASS Number	Cosmos Number	Plane and Slot	Frequency Channel	Launch Date	Introduction Date	Status
784	2363	1,8	8	12/30/98	1/29/99	Operating
786	2362	1,7	7	12/30/98	1/29/99	Operating
783	2374	3,18	10	10/13/00	1/5/01	Operating
787	2375	3,17	5	10/13/00	11/4/00	Operating
788	2376	3,24	3	10/13/00	11/21/00	Operating
789	2381	1,3	12	12/1/01	1/4/02	Operating
790	2380	1,6	9	12/1/01	1/4/02	Operating
711	2382	1,5		12/1/01		

[a]All the dates are given in Moscow time (UTC + 0300).

since that time. At present there are between 32 and 35 GPS satellites in orbit. For the system to operate, a minimum of 21 must be functional at any given time.

5.4 GPS ACCURACY AND DESCRIPTION

NAVSTAR took a quantum leap forward on May 1, 2000 when the president of the United States indicated that *selective availability* (SA) would be turned off for GPS satellites (U.S. Department of Commerce, 2000). Until that time civilian GPS use provided for a minimum of ±100-mm horizontal accuracy 95 percent of the time without correction of the GPS standard signal. In the vertical plane, the accuracy was ±186 m uncorrected. Keep those figures in mind: Should SA ever be turned back on, those are the levels of accuracy that one could expect using GPS civilian code. The current *SPS Performance Standard* published by the U.S. government is ±37 m horizontal and 77 m vertical 95 percent of the time. With SA turned off, civilian use of GPS is often much better and regularly achieves a horizontal accuracy of about 20 m while having a vertical accuracy of about 50 m, although local position and other factors can affect accuracy. The *circular error of probability* (CEP), a measure of accuracy of the Earth positions using GPS, is easily calculated (Wormley, 2003). It should be remembered that GPS satellites are moving continuously; therefore, Earth position accuracy changes as "visible satellites" change.

For a satellite to be visible, it must have a line of sight to Earth located by GPS handheld navigators. Many users of GPS assume that horizontal and vertical accuracy remains constant; however, it does vary minute by minute over the course of a day. These changes are due to numerous factors, including satellite position, atmospherics, user environment, multipath and equipment quality, and *occupation time,* the length of time taken to record a GPS position. This raises an important issue for integrated geotechnology users. In GIS, data about the data (metadata) are collected and maintained which describe the data: datum used, analysis methods, lineage, year, and so on. Few people provide metadata with their GPS positions, although some organizations do collect the information or have written guidelines for the operation and use of GPS receivers (British Columbia Ministry of Forests, 2001). Other organizations have developed software that is available for the creation of metadata from GPS data formats (UNAVCO, n.d.). Just as with GIS metadata, collecting GPS information and maintaining this metadata information is useful and valuable when working with integrated data sets that include GPS information. Therefore, the GIS professional values these metadata, since they provide a measure of quality as well as providing information about which GIS settings to use for imported GPS data.

In practice, GPS will provide ±37 m accuracy 95 percent of the time, but what about the other 5 percent? For some applications (i.e., general hiking), ±37 m may not be important, but such accuracy is extremely important for other applications, such as navigating an airplane. The U.S. Department of Transportation has determined that the most stringent requirements for aircraft landing are 4.1 m horizontal accuracy and 0.6 m vertical accuracy. As well, applications such as moving a 30-m length of farm equipment or geocoding an address require higher levels of accuracy, since a ±20-m error using large farm equipment can mean the difference between cultivating a field and cultivated roads, creeks, and other obstacles and assigning correct addresses to individual homes and apartments.

Geocoding is the process of connecting street names and addresses to coordinates. This provides the capability to query databases by either name or coordinate. There are varying levels of geocoding, and the most useful geocoding is that which can identify and pinpoint location exactly. This is not an easy task. Many geocoded databases use postal codes or zip codes. These are general codes that cover large territories and may include many houses and buildings in one area. To increase the accuracy of geocoded databases, individual streets may be assigned coordinates, although this can lead to problems, particularly where streets are very long, thus covering a larger territory. Ideally, geocodes should identify individual houses or buildings. Beyond that they might also include (x,y,z) coordinates. To geocode a house on a street would involve the street name and number. But not all street-location are single-family houses. Instead, some of these may be apartments, stores, schools, and churches. They may also vary in size and height. GPS is useful for determining each of these coordinates, but the technology itself has errors.

To compare (x,y) errors, in Figure 5.1 two sets of contours were generated for individual GPS waypoints in a GIS. The second set of waypoints were shifted in position ±20 m from the first. All elevations and the contour interval remained the same, the only difference being location—some remained in their original position whereas others shifted due to GPS error budget anomalies.

The contours clearly indicate that if GPS points have an error of ±20 m, when contours are generated using GIS, they can result in significant differences. This difference may not be as significant on flat terrain, but in undulating terrain or mountains, these differences are pronounced. Using the same data and assuming that GPS vertical accuracy is ±50 m, elevations in the z-axis are represented for the first set of

Figure 5.1 GPS error comparison: contours.

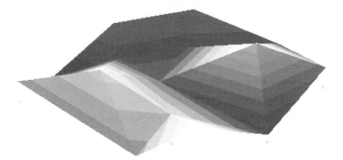

Figure 5.2 TIN created from GPS waypoints, view 1.

points (from the contours). Thus, possible elevation differences for the first set of GPS waypoints are shown in Figure 5.2. Then using the second set of waypoints, the elevations plotted from the contour diagram are shown in Figure 5.3.

The differences observed are the result of GPS errors in vertical accuracy figures published for GPS without differential correction. This highlights the impact that inherent errors in GPS can have on GIS modeling of elevations. Such differences can have major ramifications when, for example, building roads, modeling soil erosion, or performing other GIS spatial analysis involving slopes and elevations. Many applications of GPS for GIS applications therefore do not depend on using standard GPS with its published accuracy. Instead, many GIS applications and studies will apply other GPS techniques that correct the data differentially, ensuring higher levels of accuracy. The removal of selective availability does not mean that high accuracy is achieved, simply that improved accuracy occurs since atmospheric-induced errors are still present. The application of standard GPS with its published accuracy does, however, have many useful applications, which are more general in nature, not requiring high levels of accuracy. General hiking, boating, field location, and city navigation are examples of such applications. When using standard GPS data, it would appear that that the greatest concern lies in applications where not only high levels of

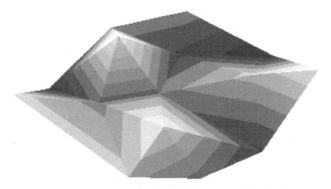

Figure 5.3 TIN created from GPS waypoints, view 2.

accuracy for navigation are involved, but where spatial analysis of information is to be conducted.

5.5 GPS SEGMENTS

GPS consists of three segments:

- *Ground control segment:* consists of five monitor stations, including Diego Garcia, Colorado Springs, Hawaii, Kwajalein, and Ascension Island, in addition to three ground antennas located on Ascension Island, Diego Garcia, and Kwajalein, as well as the master control station (MCS) located at Falcon Air Force Base in Colorado
- *User segment:* that portion of GPS the user is most familiar with—the GPS handheld navigator and its use
- *Space segment:* those elements associated with the GPS satellites themselves

The GPS system consists of a minimum number of 21 operating satellites. The U.S. GPS specifications have been published and are available online detailing the system (NAVCEN, 2001). The current GPS constellation consists of 28 block II/IIA/IIR satellites. A *constellation* represents the numbers and positions of the satellites in the sky collectively. These satellites are moving continually and circle the Earth twice a day at a range of about 20,000 km from Earth. Generally, we consider visible satellites as being available for use from the ground and collectively are termed a constellation.

To operate, each satellite has an onboard atomic clock that provides an extremely precise time base. The satellites send signals to the receiver, and the extremely precise time bases make it possible for the receiver to determine exactly how far away each satellite is. This is accomplished by calculating how long it takes for the signal to travel from the satellite to the receiver through a process called *trilateration.* Each GPS navigator stores an *almanac,* which indicates where each satellite is in its orbit at any given time. The almanac is possible because of the extremely precise orbits flown by the satellites, which are determined by the U.S. government. Consequently, when GPS satellites are being tracked, a new almanac can be downloaded to the GPS navigator, although some GPS equipment will maintain an almanac automatically if used long enough—equipment varies. The advantage to maintaining an almanac is reduced startup time. When a GPS navigator is turned on, it begins immediately to search the sky for GPS satellites. Because it does not know which satellites may be available for a given location at a particular date and time, it searches for them all, which may be a lengthy process. Having an almanac reduces this broad search for available GPS satellites and allows the GPS navigator to focus on those satellites known to be operating within specific geographic regions on Earth for the given date and time. Almanac and GPS ephemeris are not the same. *Ephemeris* refers to the orbits of the satellites themselves—how and where they move through space to maintain consistent usable orbits, thus data. Knowing the almanac and ephemeris can result in very significantly reduced startup times, saving battery power and enabling work to begin much sooner. With the decrease in SA has come improved response in timing and recovery times when GPS signals are interrupted, GPS accuracy stabilizing much more quickly. Therefore, it is always recommended that a fresh almanac be maintained when doing GPS work.

GPS operates work in the microwave band (L1, 1575.42 MHz coarse acquisition or CA-code) as well as use of a second band (L2, 1227.42 MHz). Military users have access to a much more accurate (precision code or P-code) on the L1 band. The second L2 band also results in the user being able to determine real-time ionosphere errors. These signals can pass through glass but are absorbed by water molecules (wood, heavy foliage) and reflect off concrete, steel, and rock. This means that GPS units have some trouble operating in rain forests, urban jungles, deep canyons, inside automobiles and boats, and in heavy snowfall (Bullock, 1997). The same numbers of satellites are not always visible to a user position at any given location; instead, they vary from 0 to 12 in most cases, although the number can be higher, especially atop tall mountains and other high locations.

Usually, if a GPS is held in the open at any time, it will on average find at least six satellites, although perhaps fewer or more. That difference in terms of the number of available satellites contributes to the overall positional accuracy that one can achieve, but a minimum of at least three satellites must be visible to locate a 2D position, and four satellites are required to locate a position in 3D (x,y,z) coordinates. The greater number of satellites being tracked contributes to higher levels of accuracy, although factors such as clocks, moisture, local terrain, and atmospheric effects contribute to decreasing levels of accuracy. Accurate range or distance from a position on Earth to GPS satellites determines accurate position. GPS is a time-based system. Range is determined by the time it takes for GPS code to be received at the GPS navigator. Any differences between those two times is termed *pseudorange* (PRN). If there is no time difference, the Earth position is accurate.

Larger time differences result in large position shifts from the accurate position. It is that principle that is used for selective ability, where ground control stations would purposely cause small changes in timing to GPS satellites, effectively creating the previous error of ±100 m, experienced before the ending of SA while using GPS. With the recent decrease in SA, errors are now in the range ±37 m or so. In other words, those small variations introduced in timing were accounting for almost 80 m of the previous error. That is why SA had such a pronounced effect on GPS position accuracy.

To acquire position the receiver code is shifted until maximum correlation is achieved between the two codes—that of the receiver and that of the satellite. The time magnitude of the shift is the receiver's measure of pseudorange time. Think of GPS satellites and a GPS navigator both generating strings of long codes for each satellite. The GPS codes would consist of a series of binary digits (i.e., 1001100110) from each. If at a point in time, both satellites and navigator are generating the same codes (strings), they achieve *lock* and can be used to determine position. If at one point in time the navigator must shift its code (in time) to enable it to lock, there is a time delay before lock, leading to less accurate positioning (i.e., PRN). The GPS user has no control over GPS time since that is preset by the U.S. military and based on atomic clocks. However, because the U.S. military can control the delay of the time signals, they can reintroduce timing error.

5.6 ERROR BUDGET

When we speak of GPS position errors of ± 20 m horizontal and ± 60 m vertical, we are referring to a portion of the error budget. Given that we still can achieve only about ± 37 m position accuracy with GPS, there must remain errors in time. Where do they come from? The principal remaining factors contributing to the overall error budget include clock and equipment and atmospheric and multipath factors. Therefore, some error in GPS position will always be present, although they can vary and are partially manageable. Good-quality navigators operating in conditions that reduce *multipath* and other effects can go a long way toward improving GPS position. Multipath errors are caused by the GPS signal having to travel a longer distance before reaching your GPS navigator. This is due to blockage of the path between a GPS satellite and the receiver.

Because the signal travels farther, it takes longer, therefore contributing to greater error in position. Multipath operations can be characterized differently depending on the conditions present (Mora et al., 1998). As an example, let's assume that we are walking downtown in a large city with a GPS recording a series of waypoints that form a line, commonly referred to as a *route* (for this we assume that there is differential correction). The GPS sample interval is preset to take a GPS waypoint at regular time intervals, and all that is needed once the route is initiated is to begin walking the preferred route. Moving along the path, there are some larger buildings on one side than on the other. Finally, we move to a point where we are surrounded by tall buildings, then back out into the open and end our data capture. Returning to the office, the data for the route are downloaded readily and imported into a GIS for viewing. The route does not look at all like the route walked; many points veer to the left or right, and some of them even suggest that the street width may be 90 m or more across, something we did not see while capturing the data, and wonder about.

The tall buildings have contributed to the errors in position through multipath. Signals bouncing from cement and metal surfaces have all traveled farther before arriving at the GPS navigator. The added length of those distances has in effect delayed the original times. Multipath can be reduced using choke rings surrounding the GPS antenna, potentially preventing signal from arriving at the antenna from below and below the horizontal plane. (Foresters know about multipath since it also happens in heavily canopied forests, particularly when wet, which attenuates GPS signals.) The question remains, however: What do we do about creating accurately the route that we set out to obtain? It cannot be assumed that all the errors are attributable to multipath; a portion of them will be due to atmospheric effects and time clock errors in the equipment. It is not easy to apportion the errors, nor are we really interested in doing that—all we wish for is a map of the route.

The use of other geotechnologies can become useful at this point, including base maps, aerial photos, satellite images, and GIS. First, we could look at the table and attempt to delete those sample points that simply appear very incorrect. So we begin deleting points from our data table, and for each one deleted, others appear to be outside the spacing of the remaining points. After awhile fewer and fewer points are left, and if plotted, the points do not at all represent the route traveled. This is a problem with tampering with data sets arbitrarily—one can never be sure what is real (and useful) data and what is not—and is compounded by subjective interpretations made during the weeding process.

It would be helpful to incorporate a remotely sensed image with known coordinates or an aerial photograph into a GIS as a backdrop, registering the GPS points against the backdrop. In this way, those points that appear well outside the possible route could easily be deleted. The added benefit, of course, would be that the backdrop image could be georeferenced from the GPS points and become useful for other purposes. A backdrop could also be interpreted, resulting in the addition of thematic themes in addition to the route. For this application to work, the images used for backdrops would have to be of sufficient resolution that objects are easily discernible and the route is represented in sufficient detail. The reader might by now be wondering why GPS is being used to reference the route if it is already on images? The answer is that although it is on the images, it has never been digitized for use in GIS as a theme or within a thematic layer. Now that it is, other GIS analysis may be possible with respect to the route, such as number of people who live within 10 km of the route to determine if it is a good place to set up an information booth.

5.7   DILUTION OF PRECISION

To capture a 2D position using GPS, a minimum of three satellites must be tracked. To capture a 3D position (x,y,z), a minimum of four GPS satellites must be tracked and recorded. The decision to capture GPS data in 2D or 3D is depends on the application. When only (x,y) coordinates are needed, a planar map can be constructed. When elevations are to be included in the data set, a 3D GPS position can be captured. Capturing 2D GPS data waypoints takes considerably less time in most instances, particularly in environments where four satellites are not likely to be available for tracking. *Elevation angle* is to the angle facing skyward from which a GPS navigator can view available satellites (Figure 5.4).

Assuming that we wish to capture a 2D position for X, the elevation angle in this figure is about $45°$. If we widen the elevation angle, for example to $180°$, the line would be flat, conforming to the road. If the elevation angle were narrowed to $5°$, the arrow would decrease, becoming much sharper and more pointed. The reason for changing the elevation angle is to discard satellites far off along the horizon where they must pass through more atmosphere, thus contributing more timing errors. This allows the GPS instrument to focus and cycle only through those satellites within the set elevation angle. In a densely wooded environment, the chances of multipath are increased if satellites off along the horizon are used, because the signal must pass through more trees. Alternatively, decreasing the elevation angle, to include only

Figure 5.4   GPS elevation angle.

those satellites directly above results in poorer satellite geometry, referred to as geometric dilution of precision. There are other dilutions of precision. The primary ones most GPS users will be interested in are as follows:

- *Positional dilution of precision* (PDOP): measure of overall position accuracy
- *Vertical dilution of precision* (VDOP): measure of accuracy in the z-plane
- *Horizontal dilution of precision* (HDOP): measure of accuracy in the x–y plane
- *Timing dilution of precision* (TDOP): measure of accuracy in time
- *Geometric dilution of precision* (GDOP): measure of satellite geometry

Generally speaking, PDOP values below 5 are good for performing GPS work; even lower would be better. A DOP value above 5 is indicative of contributing errors. It is wholly possible to have a high VDOP but a low HDOP. That is, the elevation coordinate may be poor while the (x,y) coordinate may be very good. Each DOP contributes to the overall final position accuracy (or inaccuracy). Where GPS information is gathered and applied, a record of DOP should be kept. This will not only form the basis for inclusion into metadata but can be useful for determining the accuracy of any recorded position. Once GPS data are downloaded, not only are these data points provided from GPS software, but the DOP measures associated with individual points are often included. A quick cycling through data points and their associated DOP values can provide important clues about the quality of the data points captured.

There are other very practical reasons for including GPS DOP measurements within data sets. They provide a time record in the event of a legal dispute, as well as providing to contractors and clients a measure of one's quality of work. In worst-case scenarios, poor DOP values associated with some applications (e.g., guiding large ships or other mission-critical applications) can have major negative repercussions. In such cases it is not unusual to set a DOP limit on the GPS navigator such that it will choose only the lowest DOP values and/or discard higher values. This can be a trade-off, since raising the threshold of usable DOP values may also result in having to wait around for usable satellites. Keep in mind that some sources of contributing errors may in fact be outside the capability of the GPS user, as they are more related to atmospherics, solar flares, multipath, or local conditions. Whichever approach is used, it must be possible to generate real-time navigation information fast enough for safe navigation.

A GPS data set can often be exported to a GIS as a comma-delineated table or DXF file. Some GPS software allows for direct transformation to particular brands of GIS upon export. Maintaining DOP values with data points has other uses in a GIS. Because a GIS has query functions, the GPS data tables can be queried not only for particular attributes but also for selected DOP values. If a GIS mapping contract calls for inclusion only of GPS values with a DOP value of 3 or less, the GIS can query exactly for those values and provide a map. Remember, DOP is not a measure of detail—it is a measure of accuracy. Metadata associated with GIS information should include GPS details since they are indicative of spatial accuracy.

5.8 DIFFERENTIAL CORRECTION

Differential GPS is a method used to improve positional accuracy with respect to standard GPS pseudoranged positions. It is sometimes referred to as *differential correction*. The GPS user must utilize a base station operation where a differential global positioning system (DGPS) is to be used to increase accuracy to within 1 m or better from the current noncorrected published accuracy of about ±20 m. There are two methods that can be used to apply DGPS and that depends on application:

- *Postprocessing DGPS* (applied after data are collected)
- *Real-time DGPS* (where corrections are used in real time)

DGPS will require careful planning and attention to interoperability between the GPS units themselves, software and hardware used to download the information. In recent years numerous efforts have been made by manufacturers to ensure high levels of compatibility and efficient data transfer protocols. It is now possible to correct errors and data by various brands of GPS. The principle behind DGPS is very simple (Figure 5.5). At any two locations, navigators and GPS receivers will receive the same errors (inaccuracy). This implies that they will record information with the same errors and timing differences. In fact, all GPS receivers in the geographical area of about 250 km will record many of the same errors, tracking the same satellites. There may be slight differences due to atmospherics, location characteristics, and the like, but in

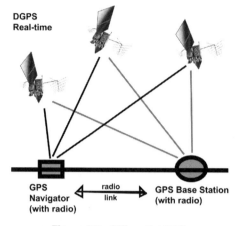

Figure 5.5 Differential GPS.

most cases all receivers will exhibit similar inaccuracy. Thus, if the user were to record all the satellite errors at one known location, a location accurately surveyed whose coordinates are precise, and then correct the data mathematically with reference to that location, all the satellite data timing errors (timing differences) would be reduced and adjusted so as to be negated.

The collection and processing of these signals is called a *base station.* Where real-time operations are involved, a need for the base station to collect and store GPS data continually decreases or becomes unnecessary, since, data are processed and transmitted in real time. In a postprocessing environment, the GPS data collected from each satellite being tracked must be stored for later mathematical correction and processing that includes the field navigator data. To perform DGPS corrections of GPS satellite data, it is necessary that both navigator and base station are receiving signals from the same satellites. Remember, to receive 3D signals requires the use of four satellites. Therefore, for DGPS to work in 3D means that both base station and navigator must be tracking and recording information from the same four satellites. If either one drops a satellite for whatever reason, it is not possible to achieve 3D correction. In this case only 2D information would be available, since a minimal number of three satellites would be tracked between both GPS receivers.

There are other limitations to DGPS. These include long baseline distances. The farther apart the navigator from the base station, the poorer the chance of tracking the same satellites (the satellites descend over the horizon). The accuracy also decreases over longer distances where atmospheric effects increase signal errors. Generally speaking, DGPS can be performed at distances up to 250 km from ground-based GPS stations and navigators. The terrain will greatly affect this distance; flatter terrain will allow for longer baselines than will rolling hills or mountains. Obviously, where units are placed in higher locations (i.e., in airplanes), there is a greater chance of receiving satellite signals. Thus, there are no hard-and-fast distances over which DGPS cannot be performed. Over longer distances the accuracy will decrease the farther the two are apart. This inaccuracy may amount to 1 to 3 m or more at greater distances. Usually, both navigators and base stations are within 250 km or so in most applications. The process of applying DGPS involves taking the errors received at the base station and processing them to derive the computed position of the base station. Since the navigators receive the same errors, it is simply a matter of applying corrections to the data collected from the field navigators. There are numerous considerations as to how this is done. If the application is in real time, the corrections must be transferred immediately. If the data are not mission critical but can be integrated later, postprocessing may be used. During real-time applications the mathematical corrections or processing for DGPS is completed and transferred immediately to the field application requiring accurate positioning. This can be accomplished by radio, demanding that both navigators and base station have an established radio link. The disadvantage to a radio link is distance, since most radios do not operate beyond 30 km or so—otherwise requiring a broadcast license. Alternatively, an uplink to a communications satellite may be used to transfer the corrections to the field navigator. In these cases, the field navigators require both a GPS system for tracking current positions and a communications satellite capability for linking to the corrections being transferred from the base station. Is one better than the other? Not necessarily, since both radio and satellite links provide similar corrections. The advantage to satellite communicating, though, is that the field GPS can be moved far greater distances than 30 km.

Operating a base station and linking to GPS also requires knowledge and hardware since radio and telecommunications equipment must be maintained. With a satellite-based communication for corrections, the user essentially rents time from a DGPS service provider whose primary business is providing DGPS corrections. One can buy corrections when they need them and avoid the processing of unnecessary files. If you are a heavy and consistent user of DGPS correction information, you may also be offered lower pricing. Wide-area differential GPS (WADGPS) consists of a series of base stations across a region providing a wide area of coverage for correction data. Depending on the distance to the nearest WADGPS station, GPS information may be corrected to higher levels of accuracy. This stands to reason since the user's position in the field may be hundreds of kilometers from the nearest WADGPS station, tracking different satellites. Wide-area augmentation systems (WAASs) are similar to WADGPS but designed for continental or international use. A WAAS consists of a series of base stations located throughout a country that calculate corrections for any positions from available data, then transfer those corrections through satellite uplink back to users, who must link to the WAAS satellite, providing corrections. Current stations are optimized for positions below 50° latitude. This means that in the most northerly and southerly latitudes, more accurate position fixes are achieved with more local WADGPS networks or by setting up one's own base station within the local region. There are several continuously operating reference stations (National Geodetic Survey, NOAA) in the United States (Figure 5.6).

In Europe, EUREF (European Reference Frame; Gartner, 1996), which is a subcommission of the International Association on Global and Regional Geodetic Networks, operates several permanent tracking stations (Figure 5.7). Plans are under way for the development of another satellite service providing European DGPS correc-

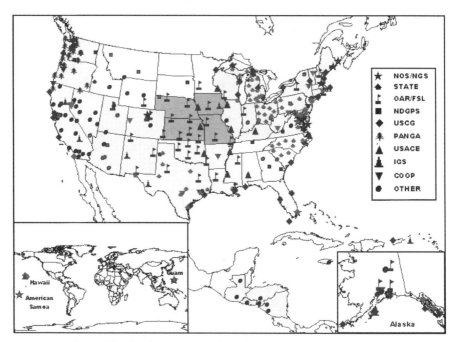

Figure 5.6   Continuously operating reference stations, January 2002.

tions incorporating both NAVSTAR and GLONASS satellites (EGNOS, 1994). In Australia, GPS data in RINEX format may be processed online and mailed back to the user with corrected positions (AUSPOS, 2002). Individual users may even use two independent receivers locally for DGPS applications. This can be accomplished in the following way, provided that the GPS navigators being used are capable of recording satellite information and have sufficient memory. One navigator can be placed in a known location, serving as a base station and begin to record satellite information continuously while the other is operated in the field. Once the assignment is complete, the base station navigator data can be corrected to the known position, then those corrections applied in a postprocessing manner to the field GPS. If real-time information is necessary, establish a radio link between the navigators. The reader may be wondering how this relates to GIS. The accuracy of the waypoints taken with GPS directly affects the quality of work achievable with GIS. Even simple things such as determining the density of bird nesting locations, proximity to roads, and similar analysis can be greatly skewed when comparing GPS point data with ±1-m accuracy rather than ±20-m accuracy. Many GIS users tend to import GPS data into GIS, then perform these types of spatial analysis believing that they are correct without questioning the accuracy of the GPS measurements—yet another reason to ask for a GPS metadata.

The correct name is RTCM-104, Version 2 or 2.1. The acronym stands for "radio technical committee—marine" and is the name of the committee that governs stan-

Figure 5.7 European reference frame.

dards for passing data between different equipment used in the marine electronics industry. This group developed a standard format for sending differential correction data to a GPS receiver. The actual format is complex and lengthy, but it contains three main elements:

- Time of the measurement at the base station
- Measured range errors (corrections) for every satellite in view at that base station
- Range error rate for every satellite in view

Range error rate allows a GPS receiver to continue to use the last correction it received until a new one arrives. The time it uses that last correction is known as the *age of the data*. A reasonable rule of thumb is that in about 12 to 15 seconds the error will have grown to exceed 1 m. After that, growth may be very rapid. GPS receivers have a settable parameter as to how long to let this age increase before going to a nondifferential mode. It is recommended that this be set at no less than 15 seconds and no longer than 30 seconds.

5.9   GPS AND GIS SURFACES

When a series of GPS points are taken for the purposes of creating a surface in GIS, those points may be discrete or nondiscrete. (For additional information on surfaces, see Chapter 4.) *Discrete points* consist of GPS locations taken wherever the user decides. *Nondiscrete points,* on the other hand, form a regular grid and are spaced evenly. Either discrete or nondiscrete GPS points may be used to generate a surface. Surfaces are generated using a GIS, and there are several methods for generating a surface, each using interpolation to estimate surface values that lie between sample points (Bergh and Lofstrom, 1976). Some of the more common interpolation methods used in GIS for generating such surfaces are:

- Inverse distance weighting
- Trend analysis
- Triangulated irregular network (TIN)
- Kriging
- Splining

No single method is suitable for all applications; instead, each offers advantages and disadvantages with respect to application. Inverse distance weighting weights points, giving those closer to the actual GPS sample location higher weighting than those farther away. Where there are no sample points taken, values are weighted by distance from the nearest point, which may in some cases be a considerable distance. Trend analysis results in a smoothing of data points and provides an indication of the overall trend of the surface. Sharp rises and steep valley bottom positions are averaged using trend analysis; subsequently, their absolute position may be altered significantly depending on nearby elevations.

Triangulated irregular networks are used to create continuous surfaces from vector data (Peucker et al., 1978). Individual GPS sample points become nodes when using a TIN, which then determines other nearby nodes forming triangles between series of nodes. TIN will also maintain *breaklines,* naturally occurring boundaries such as roads and paths and other entities conforming to vector topology. Because TIN conforms to database topology (i.e., GPS data tables), there may be sharp changes in the surface generated since the surface is composed of numerous faces consisting of triangles that connect in straight lines from node to node.

Kriging is used to build surfaces and is based on regional variable theory, which assumes that variation within the surface can be analyzed statistically and that homogeneous variation exists throughout the surface (Stein, 1999). Any variation is represented using a semivariogram, which, in turn, is a measure of correlation between estimated and observed points. Weightings are assigned through use of the semivariogram. Unique to kriging is that as more and more points are clustered together, their weighting becomes more aligned with the average of these points, or homogeneous, as mentioned. Where kriging has difficulty is when the sample points are not evenly distributed and become more and more irregular, effectively resulting in decreased correlation between points and where there are large surface changes (i.e., elevation changes—observable values). The application of the correct semivariogram is critical for developing an accurate surface.

Splining is an interpolation method that selects intermediary values between sample points, then estimates the new surface values generating a new surface. Using this method may result in peaks and valleys or other sharp changes in landscape not being represented in the generated surface. The density of GPS samples taken can be directly related to the interpolation method chosen for use in GIS when and if interpolation techniques are used to generate surfaces. The fewer the number of GPS sample points taken, the more positions or data values that will need to be estimated or interpolated. Even if large numbers of GPS data points are taken, each interpolation technique may interpolate the GPS points differently, resulting in differing surfaces. Clearly, the more points that can be sampled with GPS will result in surfaces that are more accurately represented. Users of GPS consider sample density, but it remains a function of cost and also access. If a GPS user cannot gain access to terrain or environments physically, interpolation provides a means to estimate a value for these points. Often, laser technologies can be used in these cases to acquire the necessary data. From a known GPS position the laser distance and azimuth are computed to derive the inaccessible position.

5.10 STATIC AND DYNAMIC SAMPLING

Static positioning is a process whereby a GPS navigator remains at a point continuously recording its position—averaging. The length of time taken at the sample point is called the *occupation time*. Longer occupation times result in higher accuracy since a number of samples are averaged to obtain one position. In fact, very accurate positions can be located using standard pseudorange GPS if a location is occupied and sampled continuously for long periods of time (i.e., 1 hour). The occupation time may vary depending on such factors as accuracy requirements, satellite geometry, type of receiver, length of baseline, and atmospheric affects (Beutler et al., 1989).

Usually, the GPS navigator occupies a position for a minimum of 15 seconds, taking a number of readings over that time, averaging them, then computing the final position. In British Columbia, Canada, operational procedures for forest resource surveys stipulate a minimum of 15 samples collected over a 30-second period for standard position locations. British Columbia, high-precision static GPS sampling would be the collection of 50 positions over a 150-second time period. Other jurisdictions will have different requirements, and users are advised to consult local survey associations and GPS agencies regarding such requirements. Static positioning is usually chosen where higher accuracy is required. However, even though occupation times may be increased, if few satellites are visible or other environmental factors are affecting signals, longer occupation times may be required. In some cases this may result in visiting the site at another time, when conditions are more suitable to data collection. There is a trade-off when static sampling since the longer time spent sampling individual points results in longer durations until the project is complete. *Kinematic sampling* (RTK) refers to collecting GPS data while moving and sometimes is called *dynamic traversing.* A real-time link may or may not be used with kinematic sampling, again, depending on the type of application, accuracy required, and whether or not the application is mission critical. RTK sampling is most commonly found in transportation applications where vehicles are moving, as well as in mission-critical applications involving emergency vehicles as well as for business delivery vehicles using GPS coupled to network analysis. It is also used where trails and roads are being mapped or where line data may be useful. That is not to say that a line cannot be made from a series of static points, but a GPS will build the topology as the route is traversed.

It should be kept in mind that the same environmental factors—atmosphere, multipath, and a multitude of others, including the navigator itself—will affect the final accuracy, just as in static sampling. Most GPS navigators are capable of setting *sampling interval time,* the interval between which each GPS data point is sampled. The longer the sampling interval when in kinematic mode, the greater the chance of losing valuable information upon the landscape. Imagine walking around a block with a GPS taking a sample once every 3 minutes. Even at a slow walking pace, the corners would be cut from the final data, resulting in a rounding of the corners. If you walk fast enough, you may find yourself back where you started with only one point being collected.

Ideally, the navigator would be held on the corners until at least one sample were taken before continuing to walk. Driving a truck, all-terrain vehicle, or motorboat, or moving with a GPS faster and farther than the sample interval, will result in loss of recording details. There are two ways around this: Either move more slowly or increase the sampling interval on the GPS. Generally speaking, the decision to switch from static sampling to dynamic sampling involves consideration of the following factors:

- Amount of memory available in the GPS navigator
- Level of accuracy needed (static sampling results are higher)
- Project completion dates and time lines
- Availability of a base station for continuous correction
- Equipment leasing and rental costs
- Costs of housing and transportation of employees if spending longer times in the field
- The application (e.g., mission criticality)

The time at which a GPS navigator actually collects data, called the *epoch,* is recorded in seconds. When processing data we sometimes run into cases where navigator and base station epochs do not match in time, called *cycle slips.* In these cases the processing computer realigns the timing epochs and restarts mathematical processing of data for time alignment, any data recorded during a "slip" being lost. Usually, there are few cycle slips in GPS data.

5.11 MISSION PLANNING

GPS mission planning is useful for determining when to be in the field based on the availability of satellites, but it is also the initial point where GPS truly integrates with GIS. It is through the development of data tables in GPS, assignment of objects, and their corresponding attributes along with values in mission planning which strongly affect future GIS use. The mission planning portion of GPS has numerous possibilities, ranging from determination of satellite orbits to potential DOP levels, sampling intervals, and on to data table design. Just as the name implies, mission planning provides the GPS user with the opportunity to preplan the work schedule based on technicalities and spatial theory. Most GPSs come with software that provides mission planning functions. Beginning with base station operations, there are four modes:

- *Tracking mode.* In GPS tracking mode the GPS base station ought to be able to track 6 to 12 or more satellites in a *multiplexing* fashion. Multiplexing means that all satellites are tracked at the same time rather than the GPS sequencing through them in a cyclical manner. Orientation of the satellite receiver will affect the numbers of satellites visible. The base station antenna should be located at a position with an unobstructed view that is free of interference, and usually at some height.

- *Integration and storage mode.* Most GPS base station software has the ability to collect raw satellite data and store the information in predetermined data files. Usually, data are used or sold for 1-hour time blocks. Base station software can have the ability to begin automatic startup and shutdown by either time or file size. Large-capacity hard drives are necessary to store the voluminous data files of raw information and later postprocessing if that method of differential correction is being used. Be forewarned, GPS raw data files can become very large, particularly when operating continuously.

- *Processing mode.* GPS base station software must be able to integrate and process data derived both from the navigators and the base station in a quick and efficient manner. The computer running the base station software should be fast enough not only to accept and store the data from 6 to 12 or more satellites but also quick enough to perform the needed mathematical calculations required in real-time differential global positioning system (DGPS) corrections. This means that telecommunication capability must be achievable while data are being processed.

- *Telecommunications mode.* A primary function of the GPS base station is to transfer and exchange information with radios, satellites, and/or other computers on a network. Most computers have network capabilities, and this is not usually a problem. One common format for data exchange of GPS information is *receiver independent exchange format* (RINEX). Many brands of equipment now include transformation capability to allow for the sharing of GPS data files.

Since real-time base station operations that do not involve storing raw data transmit corrections simultaneously, there is less need to set up FTP servers and manage large data files of raw data. If the work is mission critical or being performed at a distant location involving travel time and associated costs, there will be a need to ensure that the base station operates continuously without undue interruption. Earlier there was mention of cycle slips. In base station operations, many of these can be avoided using a computer that has a fast processor, since delays in processing corrections due to floating-point calculations can significantly affect incoming data. Having said that, any Pentium-based computer should be able to handle processing requirements. Often GPS users are surprised when they locate their base station antenna to find suddenly that they are tracking 10 or even 12 satellites at one time, whereas their field GPS equipment tracks perhaps only 4 to 7 satellites regularly.

Open exposure and a clear line of sight to satellites vastly improve the ability to track satellites. When a GPS base station antenna appears to be tracking 3 to 7 satellites, consideration should be given to changing the location, keeping in mind that a minimum of four satellites are needed for 3D coordinates. The objective, then, is to orient the antenna in such a way that the maximum numbers of satellites can be observed at any given time. Through the use of mission planning software and based on updated almanacs, future positions of GPS satellites can be ascertained for anticipated fieldwork locations. For this portion of the software to work, however, requires that updated almanacs be maintained, since the positions are based on the latest constellation information. Knowing beforehand the positions of satellites and the area in which the field navigators will be functioning allows for travel planning and determination of the best times to be in the field.

Figure 5.8 shows the satellite availability for both NAVSTAR and GLONASS satellite constellations. There have been many cases were GPS navigators were blamed for not tracking satellites, when in fact, few were visible at the time. As mentioned previously, the satellites move all day long and sometimes yield better results for certain localities than for others. The mission planning software will also provide an indication of DOP levels. This is very useful to know about beforehand, particularly if a contract or need exists for data that should have a DOP within a specified range. Alternatively, knowing that the DOP levels will be higher during a certain portion of the day, the user can adjust sampling strategy to include more static sampling.

Databases developed in GPS software can be downloaded to a GPS navigator via a computer and the RS-232 port on the navigator. Alternatively, data tables may be designed with that capability within GPS units. The value of this capability is very important to the GIS professional, since data tables drive GIS. There are two very good reasons for developing data tables, then downloading them, which have a strong effect on later GIS performance:

- Logical consistency is maintained. All objects, attributes, and values are denoted similarly, and topology conforms to point, line, and polygons by label.
- Data tables are critical to training and GPS use in the field, including interpretation guidelines and occupation times.

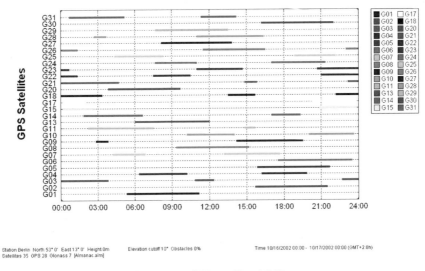

Figure 5.8 GPS satellite visibility.

Logical consistency provides the means both to understand spatial data structure and to query the data; for example, trees will be points, fire hydrants will not be lines, and polygons of vegetation cannot have fire hydrants. When consistent in querying trees, only trees will be shown, not fire hydrants. Random collection of data points results in labeling objects, attributes, and values irregularly. After a few hours the labels begin to change and the value ranges vary slightly as a result of natural human error. This happens particularly when several objects, attributes, and values are being added.

The results of randomness can become very problematic when a project entails two or more GPS operators at the same time and merging of the final data. In short, the database structure becomes corrupted and illogical and after downloading, presents a problem when applying GIS queries. By downloading database structures to GPS navigators, the structure is maintained. The operator need only press the correct buttons to enter data. Further, sampling rates, elevation angles, and occupation times for lines, points, and polygons can be predetermined and also downloaded, ensuring similar data gathering. There is not much merit in having one operator perform static sampling using a 3-second occupation time while other operators perform the same operation for 45 minutes—obviously, one set of data will be more accurate than the other. This will show up in the GIS tables by variations in overall positions and DOP levels between measurements.

As an additional consideration, when several users of GPS are working on the same project, there is a need to ensure that they are making similar interpretations when attributes and, particularly, their associated values are recorded. A red pine tree cannot be a white pine tree, and a shade of blue can have only one of four values, not one of 16 values. This means that operators are trained prior to using the GPS navigator to ensure they all understand the sampling procedures and the interpretative methodology in use. Once the GPS data structure is set and these values assigned, ideally all personnel will be operating and recording with similar consistency and conformity. Without such planning, bias may enter into a data set.

To provide an example of how quickly GPS data can become nonuseful in GIS, a GPS operator was waypoint sampling three blocks and two roads in the center of a city. The map (Figure 5.9) obviously does not show those blocks or the streets. Instead, there are a series of dots. What happened? In this case the operator collected all the GPS data as point data. Roads are usually collected as lines and blocks as polygons. A series of points can be transformed into a line, and lines closing upon themselves form blocks. The waypoints are accurate. The problem is in being able to discern roads from blocks. The waypoints are entered into a GIS and queried from the database for roads and blocks.

Now all waypoints are shown and all polygons are shown (Figure 5.10), but we queried for roads only in the GIS query. Let's try this again, this time asking only for blocks. The result of the query shows the blocks but also some of the points taken from the road (Figure 5.11). Why is this so complicated? The answer to this lies in the data structure and how the GPS points were collected. In reality, all the points were input separately, but those for the polygons had vertices joining the points to form polygons—to look like blocks. Most were labeled as numbers, but a few were labeled incorrectly as roads. Some of the points for the roads were labeled as numbers, but not all—a few were labeled as roads and some as blocks. Thus, when a query was made for roads, it included all points with roads as the label. This provides an example of how GPS programming relates directly to GIS.

Needless to say, any analysis of this information in GIS is going to provide incorrect answers. These data should have all points for the roads as lines and labeled as roads—either one, two, or possibly three, since the road branched. All the points in

Figure 5.9 GPS urban sampling.

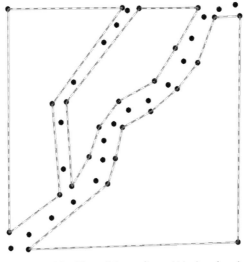

Figure 5.10   Waypoints: roads and blocks, view 1.

the blocks should be connected and labeled as block one, two, or three. Neither the roads nor the blocks are labeled consistently. Therefore, the thematic layer cannot be queried properly for use in GIS. How can this be corrected and avoided in the future? The only way to correct the data structure is to go into the data tables and label the points properly. This can be done in either GIS or using a database software program.

That is simple for this small data table, but what if you had 10,000 such points? It is clear that that would not then be feasible, which is why a GPS data structure or dic-

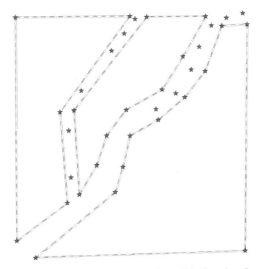

Figure 5.11   Waypoints: roads and blocks, view 2.

tionary needs to be programmed with some thought about how the GPS data are to be analyzed prior to collecting the data. Such GPS data dictionary planning can be tested beforehand by running a few queries and viewing the results. Once the database structure is optimized, it can be downloaded into a handheld GPS navigator. This is another reason why location-based linked servicing and direct access to data tables are useful. Using a mobile device, the user (client) can access the remote (server) database and simply download the data structure currently assigned to the database.

5.12   GPS VECTOR AND RASTER MODELS

GPS data can be structured similar to GIS data using a *vector model* of points, lines, and polygons. The smallest identifiable unit then is the point. A series of points form a line, and a line closing upon itself is a polygon. GPSs are designed to collect points, lines, and areas or polygons. In addition, most GPS software used for the downloading of navigators has only a rudimentary ability to display vector information as points, lines, or polygons. Subsequently, GPS points that are downloaded using GPS software and displayed can be queried, although spatial operators and more complicated spatial analysis remain in GIS. GPS is about the vector model in practice and use— it interfaces vector-based GIS quite effectively. Points, lines, and polygons exported to GIS directly from GPS can form data tables for immediate GIS analysis. The points, lines, and polygons can be queried in numerous ways, depending on how the data points are labeled.

The user of GPS software can often change data points, lines, or polygons and re-label if desired in the GPS software. This is a very tedious task and opens the door to increasing human errors in the data set. Each time GPS data are recorded, a new project or job begins. Alternatively, the GPS user may decide to append one set of data to another. This can happen for numerous reasons, including the need to stop for the evening or to take a lunch break. When a GPS navigator is restarted it will most probably ask whether or not the project is to be appended to another. Failing that, individual projects can usually be downloaded and then merged together in software. A key advantage of GPS is that it can be used to georeference aerial or satellite images, provided, of course, that the scale of those images is sufficiently small to enable the identification of points that meet the accuracy of GPS points ($<$ 1 m) collected.

Images are raster models, and as noted previously, a raster may represent distances greater than 1 m, depending on the scale. In these cases the GPS data must consider the scale of the image during any attempt to georeference a point upon them. This raises a question as to the sense of going to all that trouble to locate a GPS point accurately if it can only be represented on the image as 20 m (1 : 20,000). Therein lies one of the problems with mixing vector models with raster model spatial information. Have you ever wondered why edge matching is a problem when two DEM sheets are laid side by side? In some cases GPS was used to georeference the images, and since a pixel can overlap 20 m or so, the edges sometimes do not line up. When images are georeferenced, the entire sheet is georeferenced based on ground control points generated with GPS. In our GIS and cartography chapters, we discuss models of representation in fuller detail. In summary, GPS mission planning is important to GIS. How GPS data are to be captured directly relates to whether or not GIS analysis can be achieved in a robust manner. The quality of data transferred to GIS can be managed through observation of satellite DOP levels and satellite availability using mission planning software. Where several users are collecting similar data, training that involves standardized procedures and methods will lead to decreased propagation of errors and erroneous results.

EXERCISES

5.1. What three types of data does a GPS navigator collect?

5.2. Discuss the importance of mission planning, and identify three of its advantages.

5.3. Explain the process of differential GPS using both the real-time and post-processing methods.

5.4. What is a continuously operating reference station? How does it work?

5.5. Name and describe the three GPS segments.

5.6. Inverse distance weighting is often used in building a DEM. Assuming that GPS data for the DEM have an error of ± 25 m, in which way would they affect any DEM generated?

5.7. Name the primary sources of GPS errors and how they can be managed.

5.8. Your job is to walk 5 km, then turn around and walk back to the starting point in a thickly wooded area using a standard positioning service (SPS) with no differential correction. How close do you think you will be? Explain your answer.

5.9. Discuss issues relating to georeferencing an aerial photograph using GPS.

5.10. GPSs are capable of using UTM or latitude/longitude coordinates. Assume that you are in the middle of a project collecting data, then switch from one to the other coordinate system. What effect will this have on your data? Explain.

5.11. Given that GPS is so closely interfaced to GIS, what approach would you take to collecting GPS data about stream flow, stream depth, and width of a watershed? Why?

6

GEOTECHNOLOGY
INTEGRATION

6.1 INTRODUCTION

Currently, there is ongoing discussion internationally about the value of three-dimensional city models. These models often include draping satellite images over a DEM. Although they are useful for presentation, their value for analysis is limited. True 3D city models include topology that allows for GIS analysis in the z-axis (height). The value of a true 3D city model is significant, as it can be used for applications such as emergency services, environmental contamination, demographics, planning, and construction as well as all other applications where height is important. City models are only just beginning to become available due to a lack of 3D digital information in most cities and the large amount of work involved in constructing databases that use 3D topology. Several geotechnologies useful for collecting spatial data efficiently in cities were not available until recently. Moreover, the spatial information in cities is usually divided among the various branches of local government, necessitating new approaches to dealing with the integration of spatial information between separate organizations. Integration can therefore be seen as involving not only technology but also organizational structure.

Over the years, several people studying sociology have voiced concerns about spatial science relative to sociology. In one initiative (Initiative I-19; NCGIA, n.d.) issues were discussed relating to GIS functionality and how spatial information is represented within a social context. Concepts of representation models (raster and vector data models) were still being developed at that time and they still are. How populations and human interactions are mapped and analyzed based on those GIS structures and functionality gives rise to some concerns. Do political attitudes change sharply across cities as GIS boundaries indicate? Probably not. Does the level of educational support funding influence a region only in one GIS raster cell? Probably not. Is the drinking water quality good on one side of the street but not the other because of how the GIS is handling geocoding? Probably not. The application of GIS means the application of vector or raster models in most cases and the associated artifacts that influence the interpretation of spatial information.

In the late 1970s, triangulated irregular networks (TINs) were developed. One TIN research group located at Simon Fraser University, Canada, led by Thomas Poiker, was instrumental in the development of TINs (Mark, 1997). TINs provide a means to represent 3D topology. Triangles were constructed between points or nodes, resulting in a continuous surface. A small consulting company, Environmental Systems Research Institute (ESRI), was initiated in the early 1970s. That company focused on creating applications with a view to developing a software package suitable for a large number of users. By 1981 this software was called Arc/Info and based on concepts similar to those of DIME, using points, lines, and polygons to build a working topological structure. Others followed suit, notably Intergraph Corporation, which was also interested in developing hardware that provided the speed and capability for such applications, whose algorithms later became the instruction sets for Pentium computer chips, together with GeoMedia mapping products. Others in Europe were also developing GIS software (UCGIS, 1997). Computer-assisted design was also evolving during this time, as could be expected, since survey mapping constituted the largest available and identifiable group that might benefit from computer mapping. Agriculture and forestry were considered to be the disciplines with the strongest interests in GIS at that time because of the need to manage large tracts of land and due to the fact that the initial systems were evolving from the Canada land inventory system.

By the early 1980s a major convergence occurred that would bring GIS into the mainstream. Until this time, GIS was operating primarily under a UNIX computing environment. UNIX was free but took a lot of effort to learn, and X-Windows had not yet been developed. Therefore, computer users were required to learn, program, and use UNIX through the command line,—keying text rather than pointing a mouse. Although effective, UNIX required that the user understood not only GIS but also command line instructions and the operating system. The Geographic Raster Analysis Spatial System (GRASS) evolved in the mid-1980s from UNIX. GRASS is currently used worldwide, and many users continue freely to contribute scripts and programming toward its extended development.

As the 1980s ended, UNIX slowly gave way to the less complicated and easier to implement Windows environment for many GIS users. This meant that users did not have to remember command line instructions and could focus on GIS analysis in a more fully integrated manner. Graphic displays remained in the range 300 to 600 dots per inch for most of the decade but were improving steadily. Several stand-alone database products were developed, and digitizing and scanning equipment became more readily available. Several converging technologies in the 1980s contributed to the development of GIS. The 1980s can be considered the GIS development decade, whereas the 1990s shifted toward applications and integration, extending toward modeling and mobile applications as the 1990s ended (Figure 6.1).

It is through the development of GIS that it is most readily defined. The early emphasis on the technical aspects of the system and its evolution from computer graphics and databases led to a focus in database structures, encoding, and to a lesser ex-

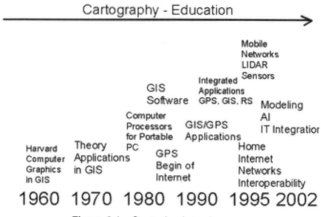

Figure 6.1   Geotechnology changes.

tent, graphics. Hardware and software permitted the storage, management, and analysis of spatial information. Prices dropped and home and business GIS use soared. Improved graphical displays and rendering tools resulted in clearer and crisper displays, and the Internet led to the ability to transfer files between users. The ability to share information through networks is often overlooked in GIS history. This is due to the fact that interoperability or shared communication and integration did not become important until the late 1990s. Data sharing through interoperability provides the easiest method of populating a local computer for GIS use. Data are downloaded from external computers through computer networks and readily imported into local computers for GIS use. Previously, most GIS information did not fit on a floppy disk, precluding data sharing. Also, network and interoperability did not exist until more recently.

Cartographic and geographic theory and spatial analysis concepts have expanded through each of these developments. A challenge for GIS has been to integrate not only technology but also spatial theory and cartographic principles. Many have wondered: Is GIS a tool, or is GIS a science? (Wright et al., 1997). Further, if it is a system, what aspects of science can be applied to GIS? These are interesting questions which no doubt affect employment job definitions and hiring practices. The quality or value of the spatial information rendered and the results produced remain dependent on the usefulness of the information for solving the problem at hand.

In this context, quality is directly related to understanding geography, cartography, and spatial analysis. It also depends on developing methods and procedures for asking and resolving spatial questions, provided that the spatial data are reasonably accurate. Consequently, GIS can be seen as both tool and theory, as mentioned in Chapter 4. Yes—one can apply GIS technology and derive a very inaccurate and nonmeaningful result: *GIS technology focused.* Alternatively, one does not need a GIS (although it is easier) to perform spatial analysis for real problems: *science focused.* Using this approach spatial statistics and modeling can be performed without using a GIS, relying on spatial theory and modeling knowledge. Therefore, the science perspective is not only about GIS technology but also about discipline-specific knowledge and spatial theory.

6.2 RESEARCH PARADIGM

An example of the research paradigm can be seen when considering GIS use within scientific organizations for research purposes. Most researchers acquire spatial information, develop models, and run their models using high-capacity computers—supercomputers. Examples of this include such areas as climate modeling, plant and forest modeling, and hydrological and atmospheric modeling. Usually, researchers are more interested in the model output numerically, seldom viewing the output graphically. Fortran is the computing language of choice by many modelers. Comparison between individual outputs is then accomplished numerically, often at the process level. At the same time, most researchers will indicate that one of the largest hurdles they are faced with is communicating their modeling initiatives and results to others. By and large, researchers are not interested in graphics, visualization, and production of hardcopy maps. They are, instead, interested in how variables interact with each other and the conditions that influence those interactions.

At the same time, research proposals and initiatives are formulated with a view to real-world problems, attempting to understand and describe how processes work and how they are affected. These results are then often used for the basis of formulating policy initiatives—affecting real people. The difficulty becomes communication of scientific research and phenomena: often termed *extension, applied extension,* or *technology transfer.* The goal of technology transfer is to move research information from the research facility into the public domain. GIS can provide the means to do this, although many organizations have not fully embraced the capability of GIS to do so. GIS can be coupled to visualization provided that one means to achieve this due to the fact that visualization can be linked to GIS and serves an exploratory and communication function. Where the science/GIS interface appears to break down is in translating modeled results into a format that GIS can readily utilize. The output of scientific models is often complex and difficult to represent visually. This is due to a number of factors:

- Time steps (run over long periods of time)
- Complexity (processes involve numerous variables)
- Communication (determining strategic viewpoints)
- Interpretation (what it means)
- Differences in human conceptualization

Many scientific models are investigated for large areas: regionally. They seek to determine outcomes at points in time well into the future: for forestry applications, perhaps 100 to 500 years in some cases. How does one go about representing 100 years of change? At what point does a viewer of the information become saturated and confused? How many variables can a user of modeled information absorb at one time, and what is the most effective way to communicate those changing variables? Finally, upon viewing the modeled output, how does one interpret the results? Clearly, the transition from scientific results in numerical form to easily understood information pieces is more difficult to interpret spatially for the average citizen and policymaker.

This suggests a need to engage nonscientists in the delivery of results through higher levels of interaction. One method to achieve that would be through the use of GIS, particularly enabling the viewer to formulate their own questions and pose "what if" scenarios. This does not exist presently, except in very specialized prototype examples, although it is common in research institutions and programs.

Ideally, individuals ought to be able to connect scientific results into a medium that allows them to conceptualize the processes, to ask questions, and to present a visualized result. From this standpoint, individuals are then able to determine how their current farming operations, forestry practices, social indicators, and other events relate to science and how they affect and influence model variables. This would then empower people to participate more effectively in regional decision-making processes as well as participating more effectively and knowledgeably in the construction and development of regional policies. Since access and computing speeds are increasing, the possibility of running scientific models in real time and rendering their results visually back to the public and communities will become increasingly more commonplace.

Scientific research and discovery is ongoing in the discipline of GIS itself. New spatial technologies are being developed that provide additional functionality in several fields (i.e., engineering, visualization, databases, nanotechnology, spatial analysis, modeling, and artificial intelligence). The highest-quality, most efficient analysis that can be shared, however, comes from applying GIS science and GIS technology together. It is for this reason that there is increasing focus on sharing spatial information between individuals and organizations. This results in the development of higher levels of interoperability or connectedness between hardware platforms and spatial data formats.

Over the years, a number of people have attempted to define GIS (see also Chapter 4—GIS and Goodchild, 1997):

> A geographic information system is a facility for preparing, presenting, and interpreting facts that pertain to the surface of the earth. This is a broad definition...a considerably narrower definition, however, is more often employed. In common parlance, a geographic information system or GIS is a configuration of computer hardware and software specifically designed for the acquisition, maintenance, and use of cartographic data (Tomlin, 1990).

> A geographic information system (GIS) is an information system that is designed to work with data referenced by spatial or geographic coordinates. In other words, a GIS is both a database system with specific capabilities for spatially-reference data, as well [as] a set of operations for working with data.... In a sense, a GIS may be thought of as a higher-order map (Star and Estes, 1990).

> A powerful set of tools for storing and retrieving at will, transforming and displaying spatial data from the real world for a particular set of purposes (Burrough, 1986, p. 11).

> The ability of a GIS to analyze spatial data is frequently seen as a key element in its definition and has often been used as a characteristic which distinguishes the GIS from systems whose primary function is map production (Goodchild, 1988).

The functions of retrieving data, storing and managing data, transformation of that information, and analysis and display appear to be the elements common to GIS.

6.3 GIS AND IT

More recently, we are witnessing the expansion of spatial technologies into the area of Web services. This has resulted in a fundamental shift from the desktop to the organization and other organizations for the purpose of sharing spatial information. It is not uncommon to hear the phrase "GIS is IT" or the reverse. In some measure it has been mobile services that have propelled GIS into the IT arena, due largely to the need for telecommunications capability.

The IT relationship to GIS is most closely related to the delivery and distribution of spatial information. It involves a connection between servers and clients, the programming of map servers, and other functions related to interconnectivity and interoperability—one system to the next. We make a distinction between GIS or geotechnology science and IT, seeing spatial technologies more aligned with the primary definitions of GIS, which are the capture, management, analysis, and representation of spatial information. Functionally, the integration of GIS with IT poses interesting issues, due to the fact that people working with spatial information increasingly find themselves dwelling on IT factors relating to connectivity. To achieve higher levels of connectivity, whether it is to a database, spreadsheet, or another server, requires increasing attention to the IT functions of interoperability and programming associated with achieving such connections.

There can be no denial that IT is related to GIS and that it is necessary to meet the challenges of providing Web-based services. However, difficulties arise due to the fact that most GIS people are not programmers or network administrators. Conversely, most IT personnel are not familiar, for example, with spatial analysis, operators, or cartography. While providing the mechanism for delivery of spatial information, the blend of GIS with IT is analogous to car owners having to know the details of auto mechanics because they drive cars. This has posed unique concerns for spatial information personnel and the defining of their positions.

6.4 JOB DESCRIPTIONS

There are no definitive definitions for a GIS, and users of GIS may be technically or theoretically oriented, or both. Ultimately, however, it is the quality of the product generated using GIS that becomes the truest identifier of GIS, due to the fact that quality necessarily involves technology and theory. Because the pace of GIS growth is encompassing so many geotechnologies, leading toward new job functions, even job definitions for GIS personnel are not clearly defined. It is not uncommon to find job descriptions for GIS professionals that encompass both computer programming experience and cartographic and spatial science knowledge. This is due to many factors, including the following:

- GIS originated from the database and is database focused, thus requiring a sound understanding of database functionality.
- Users of spatial information are inherently interested in the technologies used to acquire, manage, analyze, and disseminate spatial information; for example, data

collection technologies are purchased and implemented with a view to how the GIS can handle the information.

- There is increasing focus on interoperability and Web mapping, requiring a rich blend of cartographic knowledge but also knowledge in computer languages (Java, XML, DHTML, Visual Basic).
- Handling of spatial information and metadata creation necessitates that technical conversion and transformation processes of spatial information be documented.

In many organizations GIS professionals are often assigned the title of systems analyst or GIS analyst. This implies that they work with GIS spatial data systems that integrate with other types of spatial information systems. Aerial photogrammetry, remote sensing, and GPS systems may be considered professions and systems within themselves. Surveying has long been considered a profession in itself. GISs integrate information from each of these systems as well as others. At the same time it is not uncommon that workers in organizations perform GIS functions and tasks as part of their regular job description. In some cases, the trend toward higher levels of GIS integration into the business process has resulted in GIS becoming part of a broader IT group. This reflects the value of spatial information and recognizes how closely spatial information ties into the business and operating processes of many organizations.

Recent trends toward e-government revolve around sharing information within the government and providing the public with access to government services. Consequently, spatial information services are seen as part of these e-government initiatives. Higher levels of interoperability between systems and data formats permit interconnectedness between groups of individuals and organizations.

Recent focus on GIS and IT merging has evolved from the client to incorporate the remote functions that involve distribution and delivery of spatial information, notably Web mapping. The shift from client architectures to client–server architectures has undeniably resulted in GIS personnel having to become more oriented around IT terminology, processes, and functions. A Web-enabled organization must contend with understanding native languages between clients and servers. Employees are needed who understand Standard Query Language (SQL) and other vendor-specific languages which enable their native systems to interact with other remote systems. The development of job definitions and descriptions has not, however, been fully expanded upon in the realm of GIS, for a number of reasons:

- Confusion between GIS and IT functions
- Lack of identifying GIS value in the organization
- Misunderstanding of what a GIS is and does
- Nonsupport

Confusion exists between GIS and IT functions, leading to inappropriately identifying criteria for hiring GIS personnel. Most GIS personnel are not network administrators and therefore have little or no knowledge of network protocols, router implementation, IP addresses, and secure systems. These functions are usually conducted by trained IT network personnel who are capable of implementing Web services across networks within intranets and out to the Internet. The network administrator is not necessarily a Webmaster or knowledgeable about GIS and is often responsible for other network functionality, including; e-mail systems, network applications, security, FTP, and other required functionality for the entire organization. Therefore, to find a GIS specialist with network administrator capabilities is extremely rare, as is the converse. Yet many advertisements for GIS personal demand these very capabilities.

The value of GIS to an organization relates to rate of return or return on investment (ROI). There are very few, if any, well-documented studies that describe return on investment for GIS systems. While the market for GIS continues to grow at about 15 percent per year, it is more difficult to determine where and how much the GIS employee contributed to the value of the organization. This again is affected by the fact that GIS and IT have become so closely linked that it becomes harder to determine value relative to one or the other. However, the spatial data functions clearly fall within the GIS realm; consequently, their value must be identified and quantified relative to GIS functions. That is, what a GIS can do may be different from what the IT functions of organizations are capable of delivering. A well-trained GIS employee may be able to extend the value of spatial information if hired to do so, the value being in what is possible. But GIS functions are clearly separate from IT functions.

A *champion* is someone in an organization, preferably someone with some influence in the financial decision-making process, who is able to articulate the value of GIS. Champions usually understand at a broader level how GIS can fit into an organization and be used effectively to increase business, provide results, and open doors to other opportunities. Because they are also involved at the financial decision-making level, they understand the resources needed for successful GIS implementation and are able to bring people together across boundaries. This results in continued support and growth for GIS within an organization.

Strategic plans for implementation, expenditure, and monitoring of spatial services are developed, and necessary GIS personnel are clearly identified. If the organization is spatial database (query) focused, GIS employees who are strong in database functions are sought. If the organization is oriented toward the production of map products, a GIS person who is knowledgeable about cartography may be more appropriate. If the organization focuses on data acquiring, a person strong in GPS or remote sensing knowledge and GIS may be utilized. The key to identifying and constructing appropriate job definitions and classifications for GIS personnel is related to the functions of the organization. In a strict sense, the term *GIS analyst* has the connotation of someone who is capable of analyzing spatial data.

6.5   CONCEPTUAL GIS INTEGRATION

Modeling involves conceptualization about objects. Walking down the street, one can see a road, fire hydrants, trees, sidewalks, cars, birds, plants, sky, clouds, and buildings. All of these objects are perceived; some are different and some are similar. To identify them as unique objects is the basis of data modeling. Sometimes the characteristics of objects or phenomena may not be as readily apparent. A field of soil may have wet and dry areas as well as areas that are intermediate in dryness. In such a case the differences are conceptualized as being different from each other, based on wetness and position in space. Alternatively, there may be more homogeneous representation of objects or areas where observable characteristics are similar.

Instrumentation and sensor technologies can be applied to determine those differences by measuring them quantitatively. This results in discretization, a scaling of large numbers of interpretations into smaller quantifiable units that can be interpreted and recognized readily. Therefore, sensors and instrumentation play a pivotal role in GIS, due to their contribution to the discretization process. The reason for this is due to the fact that instrumentation and sensors can measure and monitor rapidly and continuously, producing extremely large spatial data files in very short periods of time. Consider the area of remote sensing. Satellite images may be taken for an area over time. Each series of images is analyzed independently, then later compared. Alternatively, data logging and sensor technologies are capable of continually measuring and monitoring changes. They often function with high sampling intervals (minutes or hours). Sensors and instrumentation capture and process information dynamically, providing scaled output (input to GIS). This may occur in real time or near real time.

Trends in automatic image analysis imitate sensor and instrumentation capabilities through integrating those steps. Therefore, a shift from post real time to near or real time is taking place in the area of spatial imaging. Mobile computing and location-based servicing (LBS) is another example of this process. A data model refers to how information is encoded for GIS use. Data models are of two types. The *vector data model* consists of coordinates (or a series) in space. Using the earlier example of walking down the street, objects can be classed into points, lines, or polygons. Series of points form a line and lines closing on themselves to form polygons. Fire hydrants, trees, posts, or manhole covers are examples of points. A location on a sidewalk is also a point, but a series of locations on a sidewalk form a route or line. If the line continues around the block back to its origin, it would form a polygon.

The GIS user must necessarily contend with the geotechnologies used to collect geospatial information as well as those, which automatically analyze, extract, and construct databases suitable for GIS use. How points, lines, and polygons or raster cells are being aggregated and delineated will affect how the information can be used in a GIS. In some cases the instrumentation and methods used to gather the original information provides an aggregated solution, then discards the original values. Often, these values are useful for other analysis and should be retained. In the case of sensors and instrumentation, digital data storage modules are used to store the information. Alternatively, if in real time, the information can be acquired and allotted to a continually appending database.

6.6   COMPLEMENTARY TECHNOLOGY

GPS, surveying equipment, and laser technologies operate to provide information in vector data model format. They generate points, lines, and polygons during operation. Sensor technologies and instrumentation are also capable of generating 3D data.

An example of this can be found using water temperature sensors. Several temperature sensors may be positioned on the surface of a lake for the purpose of measuring the surface lake temperature. The water temperature may not vary significantly at the surface. Lowering the water temperature sensor into the lake will provide temperatures by depth or in three dimensions. LIDAR is another example of a technology that is able to generate 3D spatial information. This information can be imported directly into a GIS for review.

Using LIDAR, a beam of light pulses is directed skyward. That light collides with aerosols in the air. Those it does collide with will result in a return beam back to the emitter more quickly—a measure of distance as well as concentration. These time returns can be used in GIS as the z-axis, useful for 3D modeling. In this scenario the aerosols can be plotted in 3D and the volumetric concentration determined. Visualization software is based on a vector data model. Buildings and other objects consist of coordinates that are located and positioned with respect to (x, y, z) coordinates that can be represented in 3D space. These structures consist of points, lines, and polygons. In the case of polygons, textures are applied, graphic images that represent realistic surfaces. Most digital technologies can be interfaced to GIS. Analog-to-digital converters can be used to transform analog stripchart recorders and other mechanical operating devices that require digital output.

Manual data collection in text form through interpretation is often overlooked as a means to collect geospatial information. This can be a very useful method for acquiring information that is not easily quantifiable using digital technologies. These personal interpretations can be appended to spatial data values and input manually at a later date. Such interpretations tend to be highly location specific and time dependent. Included are public surveys of attitudes, on-location quizzes, feedback forms, and mail-in responses.

An example application could be the survey of people within a region of a proposed new traffic change, changing a street from two-way to one-way traffic. In such a study, the opinions and feedback of the residents in the area are important prior to the final decision. Using GPS, many people are interviewed on the street. The locations where the interview took place are recorded using GPS, and notes are taken with respect to individual responses. Ideally, similar questions would be presented. Because we are interested in the opinions of residents, we must go to where they are. Some may be found closer to the proposed location of the change; others will be found farther away. Some may live in the region; others are commuters traveling through the region. This will provide feedback regarding the proposed change from a large number of people who live both near and distant from the location but also information about where they are likely to be affected relative to the proposed change.

6.7   RASTER AND VECTOR INTEGRATION

The *field view data model* or *raster data model* is based on a grid of cells. Each cell contains information relative to the spatial extents for the individual cell. Photogrammetry, remote sensing, digital photography, and scanning produce discrete fields, pixel by pixel. Fire hydrants can be represented using a vector or raster model as well as a building, road, or agricultural field. But the catch to this is that objects with sharp boundaries that are particularly small in size (depending on scale) are not represented as clearly or analyzed as effectively as when a vector model is used. The reason for this is, again, that vectors are based on coordinates (x, y, z) which have exact positions in three-dimensional space. Raster representation is based on cells of equal dimensions—having image coordinates relative to the extents of the cell.

Within a raster cell it may be difficult to distinguish a manhole cover from a tree if the individual cell covers 10×10 m in space. This is why the trend to higher-resolution satellite imagery is becoming much more valuable, with some satellites currently providing less then 1-m pixel resolution. This level of resolution approaches accuracies similar to those obtained with GPS equipment, the difference being that satellite imagery is raster based, whereas GPS data are vector based. Thus we are beginning to see the blending of vector- and raster-based technologies for similar resolutions. In this case the satellite imagery would have to be processed in such a way that the data are transformed into vector format before the two sets of information could be analyzed together. Alternatively, of course, the vector information could be gridded into a raster matrix.

On the other hand, a raster more effectively represents large areas with similar values. This makes it much more useful for analyzing small-scale applications where less accuracy is required. Consider the number of inhabitants for a city block. In a vector model each home and the number of people within each home could be analyzed. Comparisons could be made between homes and individuals. Alternatively, a raster model representation of the same city block might result in half the block as one pixel and the other half as another pixel. Not all people within the two-block area could be compared individually. Instead, they could be compared between the two pixels. Aerial photos and remotely sensed images consist of a series of individual cells. They are imported into a GIS as grids or rasters, usually referred to beginning in the upper-leftmost corners or sometimes the lower left corner of the image, with all rasters thereafter numbered incrementally.

The extents may vary in size from one project to another but remain similar within a project. Accordingly, raster grids will have dimensions such as 1000×1000 pixels, but any two values can be used. Provided that all raster images have similar extents, the grids can be queried together and another raster grid created. Thus they are easily compared and analyzed pixel by pixel. In cases where two raster images are being used for comparison that do not have similar resolution (pixels), one or the other image must be transformed to the other. Coarser resolutions originating from small-scale applications cannot readily be transformed to finer scale or higher resolutions. But fine-scale imagery can be transformed into coarse resolutions or small scale. Therefore, images can be transformed from fine to coarse scale but not the reverse. GPS vector information is imported as points, lines, and areas and can be placed atop a raster model (i.e., backdrop). In Figure 6.2 a landscape is provided with numerous trees present. A polygon at the beach is outlined and labeled and a data point is located at the corner of four joining cells. The raster grid itself is overlaid, indicating the relative cell size from the viewer's perceptive. If each raster cell continued down toward the lake, it becomes apparent that the bends and turns along the lake

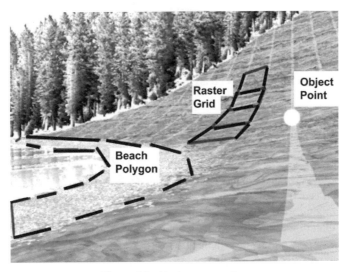

Figure 6.2   Vector upon raster.

edge would have to become represented by a pixel. It would be difficult to determine the size of the lake accurately since a pixel would probably extend beyond the edges of the polygon. To evaluate area, the numbers of pixels must be counted and totalized. The land and water may meet in the middle of a cell.

While the mind conceptualizes two separate distinct entities, the lake and the land, combining them into one raster and using that data model for analysis will result in errors. However, the lake and land can be delineated (discretized) more effectively using a vector data representation. Accordingly, GPS would be quite useful for locating the land and the lake and building the topology needed.

But what if a raster image or model already exists and we want to use it in GIS? First, the image must be evaluated for the objects and representation visible. Are they discrete or nondiscrete? Are we considering their analysis by area or by object? Are some areas more homogeneous than others? These are a few of the questions that must be asked. This is also important information to consider when purchasing a GIS. These differences in data models mean that one or the other is probably more suited for a particular type of GIS analysis or application. Purchase of a vector data model GIS when most of the work is in raster representation may not achieve the results desired, or vice versa. It used to be that one thought of either a raster- or vector-based GIS. Each model serves a particular purpose and is more useful for certain types of applications.

So why isn't everything just represented in a vector model? Although we now know that some data are better represented in one model or the other, a raster model stores information cell by cell (pixel by pixel). It takes larger amounts of disk storage for those data. Once raster analysis begins, all other thematic layers will have the same numbers of cells, the only difference being the information in them. It is during this process that a 1-Mb project suddenly spirals to become a 4-Gb monster taking up disk drive space and analysis and rendering becomes slower. Suddenly, the focus changes to issues of speed, storage, and data management. This then leads to managing, storing, and accessing data files more efficiently—data structure. How databases are structured, arranged, and designed therefore affects the operation capability of the GIS. Analysis and rendering speed are then related directly to database architecture.

6.8 DIGITIZATION

All hardcopy images can be scanned or digitized. *Digitization* is a discretization process whose objective is to identify and delineate objects on an image for inclusion in a database; it is a vector-based process. Points, lines, and areas are delineated using a small digitizing mouse that can be moved across the image. Usually, images are fixed to a small digitizing tablet or table. In operation the operator identifies features on an image or map and records them digitally with reference to their coordinates by clicking a series of keys on a digitizing mouse. This is a time-consuming process requiring patience and a steady hand. It is also the major reason why we do not see many historic maps in a vector format. Most historical and other hardcopy maps are scanned using flatbed or drum scanners.

The digitization and transformation of older information is time consuming and requires large amounts of investment. As the operator delineates features on a raster map, this sometimes results in errors, including the following:

- *Dangling nodes:* undershoots and overshoots, lines extending under or beyond polygon edges
- *Slivers:* lines not meeting adjacent to each other, leaving openings between polygons
- *Duplication:* lines, points, and polygons digitized more than once
- *Label errors and incorrect labeling:* either missing or duplicate

Much GIS software provide the means for on-screen digitization rather than the use of a digitizing tablet and mouse. Using this technique, the image to be digitized is imported into a GIS and displayed as a backdrop. The user then digitizes on the backdrop, recording keystrokes for the necessary features to be digitized. Databases are constructed during the digitization process directly into the GIS software, which can then be edited. Digitization can be accomplished in point or stream mode. In point the user locates the cursor on a point that is then recorded. This method is useful where several point features are distributed. In stream mode, the digitizing software records a point automatically as the cursor moves. The distance between recorded points can usually be preset. Stream mode is particularly effective and efficient where continuous lines and polygons are being digitized.

The cost of digitization is high. It is labor dependent and time consuming and often involves cleaning or processing the digitized data afterward, due to errors that occur during the digitization process. However, many GIS software vendors now include spatial topology filters in their products. These filters are used to process the digitized information, searching for errors such as those mentioned previously and correcting them automatically. Once digitizing is completed, the database records can be imported into a GIS and queried or viewed.

Scanning is the process of duplicating a raster image (usually, a map) using a drum or flatbed scanner. Scanning results in a duplicate image that is in raster format—without classification and with undelineated object coordinates. While vector information provides spatial data with real-world coordinates, a scanned image is a reproduction of an already existing image. To perform on-screen digitization in a GIS, a map or image must first be scanned, then imported as a backdrop. The scanning of aerial photographs can result in varying ground resolutions, depending on scale. In this case the scanned image can be imported in one of many formats (i.e., BMP, TIF, or JPG) into a GIS. We discuss scanning in more detail in Chapter 8.

6.9 FUNCTIONAL GIS INTEGRATION

GIS users are interested in GIS because of the spatial functionality that a GIS provides. If only database functionality is needed, separate database software might be a better solution. Alternatively, if only visualization is needed, a visualization package with good capability to import GIS formats might be the best way to go. But if spatial analysis is required, GIS is definitely the way to go because they are designed and optimized to include spatial functionality. Following are some of the features that a GIS includes for this purpose:

- Database creation, query, editing, import/export, and conversion
- Interfacing to digitizers, GPS, laser, and instrumentation
- Rectification, mosaic, registration, projection, and interpolation
- Spatial operations (connectivity, buffer, aggregation, and neighborhood analysis)
- Statistical analysis (statistic descriptions, classification, regression, and correlation)
- Mapping (thematic/choropleth, contour, dasymetric, charts, and 3D/4D)
- Modeling (inductive, deductive, and simulation)
- Plotting, illumination, texturing, and animation

Because of the variability between products, not all GISs have all these functions. Most GISs will, however, have many of these features, and some will have additional features. Many products related to GIS have been developed that focus on one or more of the foregoing aspects. Examples are software programs that are optimized for transformation, image analysis and remote sensing, and rendering and visualization programs capable of taking GIS output and providing high-end photo-realistic graphics.

GPS software does sometimes have database functionality. Mission planning software for GPS usually includes the ability to query waypoints, routes, and polygons.

Sometimes, extended functions in GPS software include distance operators and measurement, such as proximity analysis and buffering. It is important to look for import and export capabilities in GPS mission planning software that enable GIS linkage. Image analysis software is optimized for segmenting images and manipulating color characteristics of the image. Most image analysis software includes the capability to perform GIS operations, including encoding, thematic layering, neighborhood analysis, statistics, connectivity, and 3D surfacing. Image analysis software tends to include more functionality in the modeling functions: statements, variable assignment, matrix transformation, and dimensionality. Visualization software packages are usually able to import many common GIS formats and are designed to orient objects with respect to grids or vectors. Spatial functions, modeling, and GIS database functionality are usually not optimized in visualization software, but access and query functions are often available.

A summary of functions for GIS, GPS, and image analysis software is provided in Table 6.1. There are no definitive standards or functions that all GIS offer; instead, they tend to overlap. But GIS and image analysis software are usually the most likely to include higher levels of spatial analysis capability coupled to database and rendering functionality.

Table 6.1   Functionality of image analysis software

Function	Operation	GIS	GPS Navigator	GPS Software	Image Analysis Software	Visualization Software
Database	Import	×	×	×	×	×
	Export	×	×	×	×	×
	Query	×	×	×	×	×
	Edit	×	×	×	×	
	Transform	×		×	×	
Spatial/model	Encoding	×	×	×	×	
	Proximity	×	×	×	×	
	Interpolate	×			×	
	Project	×			×	
	Overlay	×			×	
	Algebra	×			×	
	Topographic	×		×	×	×
	Georeferencing	×		×	×	×
Model	Statistical	×			×	
	Mathematical	×			×	
	2D	×	×	×	×	×
	3D	×	×	×	×	×
	4D	×	×	×	×	×
Render	Computer	×		×	×	×
	Print	×		×	×	×
	Plot	×		×	×	×
	Visualization				×	×

The ability to link to servers is becoming more important as spatial data servers are being used increasingly rather than storing spatial information on local machines. Therefore, GIS software that includes provision for accessing data servers (i.e., Oracle, Microsoft SQL Server, IBM DB2) is very important. All scanners, printers, plotters, and digitizing equipment can usually be interfaced to UNIX or Windows operating systems. In some cases, digitizing software is included within the GIS or image analysis software.

6.10 DISK STORAGE

A single aerial photograph in color can approach 5 Mb in size, and half that size in black and white. LIDAR data files are usually quite large, and remotely sensed images, particularly in color, are large files. Visualizations are noted to have large file size requirements, particularly when animations and fly-through are used. Because data derived from geotechnologies are generated in digital format, users will often attempt to print or plot them immediately. Consideration of file sizes, color or black and white, and resolution must be considered. One way of reducing printing and plotting costs is to perform all work on the computer and rendering only when necessary. Many geotechnology users continue to have difficulty transferring their on-screen graphics to printed output. This may be due to several factors, including:

- Improper printing or plotter drivers installed
- Use of fonts not supported by the printer or plotter
- Computer software output dimensions set incorrectly
- Printer or plotter dimensions set incorrectly
- Lack of printer or plotter ink
- Device not attached to the computer

Some GPS software can be used together with GIS software running interactively. In these cases the GPS is connected to the computer and delivering GPS information to the GIS software in real time. Thus, data may be collected and stored immediately for future use, or the application may be in a real-time mode, where it can be discarded. The difference between these two approaches depends on the application. For those applications that discard information is generally navigation oriented. The GPS is used to update current position, and locations are tracked. In mapping applications data are collected and stored using similar technologies for later use in GIS and require larger disk storage space.

The processing of spatial data in a GIS can generate very large files sizes. These files often exceed the original file sizes significantly, and storage requirements for them must be planned for. The user is faced with the question of storing those files locally or in spatial data servers that serve the entire organization. Local storage is considered to be the less advantageous method within an organization where there are multiple users of the same spatial information. Through the use of spatial data servers,

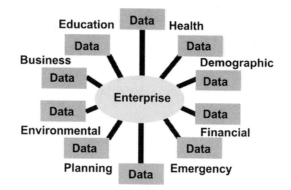

Figure 6.3   Information in the enterprise. (Courtesy of Intergraph Corporation.)

the data are stored once, regularly updated, and transactions can occur between clients and servers ensuring data integrity. The distribution of spatial data flows from one server to many users in the enterprise in this scenario (Figure 6.3).

It should be remembered that GPS navigators often have variable baud rate settings. They must be matched to the computer's baud rate settings and other telecommunications protocol. Transferring a large GPS file at 4800 baud may take a long time. Setting the highest baud rates possible is preferred.

6.11   SPATIAL INTEGRATION

Most projects involve observation, monitoring, and recording entities. *Spatial integration* refers to the integration of spatial entities using geotechnology. This could include a raster image in the case of aerial photographs or vectors in the case of GPS and LIDAR data. Numerical variables are derived from sensors and instrumentation. These pieces of information are often in different scales. The question then arises: How do I put them all together? Initially, there are a few principles about geotechnology integration that need to be considered.

- Information collected from each technology will probably have a different scale. (Aerial photographs are 1:15,000; satellite images 10 m; GPS 1 m.)
- Some information will be in raster form, whereas other information is in vector form. (GPS is vector; aerial photos and satellite images are raster; LIDAR is vector.)
- Analysis within a geotechnology use tools associated with and optimized for that technology; the same tools may not be useful between technologies. (GPS mission planning software cannot be used to analyze images; LIDAR filtering software cannot be used to thematic overlay; sensor output must be georeferenced.)
- Units between geotechnologies may vary. (GPS data are in meters, aerial photos and remotely sensed images in feet, laser measurement in centimeters.)

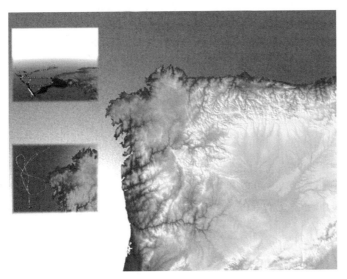

Figure 6.4 Sinking of the *Prestige*.

- There may be temporal differences; data collection both within and between geotechnologies may vary. (GPS data from differing dates, aerial photos from five years ago, sensor information by minute, remotely sensed images per two-week period.)

For some applications several types of sensors may be integrated to provide both spatial and aspatial information. An example of this is the sinking of the *Prestige* off the coast of Spain in November 2002. Carrying 70,000 metric tons of heavy fuel oil, the tanker meandered a course in the Atlantic Ocean before finally breaking in two and releasing oil which later washed along the shores. GIS, GPS, and remotely sensed images were combined to construct visualizations of that event (Figure 6.4). In this case, GTOPO DEMs provided a rough means (± 100 m) to locate the area, while GPS was accurate within meters. The resulting DEM resolution could not be adjusted downward to see landscape details; nevertheless, we were able to provide the opportunity to see the *Prestige* trail in a more generalized view.

EXERCISES

6.1. How has the definition of GIS affected the defining GIS-related job descriptions?

6.2. What is geotechnology integration?

6.3. Given that resolution varies between different geotechnologies, discuss what you consider to be a useful approach for integrating three or more technologies.

6.4. What are Web services, and how do they relate to those using GIS?

6.5. Compare scanning to digitization. Discuss the flexibility of each for GIS use.

6.6. Discuss how subtle changes in GPS accuracy can affect topographic maps and GIS raster analysis.

6.7. How does the data capture method relate to cost? Is cost the primary consideration for using one technology over another?

6.8. Assuming that you have a large dataset and have analyzed it using a GIS, how would you go about presenting the results to those unfamiliar with cartography?

6.9. Discuss what *a technology is optimized* means for a specific functionality.

6.10. How does time affect the interfacing of two or more geotechnologies?

7

SENSORS AND INSTRUMENTATION

7.1 INTRODUCTION

Instrumentation for geotechnology applications is an exciting area because these technologies are most closely linked to applications. They are designed to capture spatial information for solving spatial problems. Instrumentation in the context of spatial investigation and applications includes any instrument or sensor that may be used for measuring or acquiring spatial information related to feature characteristics and behavior, both spatially and temporally. They may be applied in real time or may continually record information that is accessed periodically. Some measurements may require higher or lower levels of accuracy and precision depending on the nature and frequency of the variable being measured. The sampling frequency and duration using instrumentation and sensors may be long or very short.

Measurements and applications of sensors and instrumentation are found in very dry desertlike environments, coastal rain forests, high altitudes, urban areas, rural areas, or in subzero environments. Sensors themselves require calibration and techniques and methods for evaluating their calibration are needed, particularly when operation in environments exposed to natural elements or located in harsh environments. The measurement of light through a forest canopy requires a device capable of measuring light incidence and level depending on specified wavelengths—a *ceptometer*. Water quality sensors may be used for measurements and monitoring in lakes or for indoor swimming pools or waste lagoons. Pressure transducers may be employed for monitoring the pouring of cement and its physical properties on construction sites or as a means to measure levels of liquids, perhaps groundwater or well depths. Snow pillows are often used to determine amounts of snowfall and penetrometers for the study of ice and soil conditions. The measurement of soil temperature uses a thermocouple or thermistor constructed of various types of metallic materials, each having different properties. The measurement of noise in an urban environment utilizes a decimeter capable of measuring loudness or specific frequencies.

These instruments can be applied to study urban noise or applied in transportation studies and have been useful for determining optimum acoustical environments. Wind speed, rainfall, wind direction, air temperature, and relative humidity together with solar radiation measurements are referred to collectively as *climate instrumentation* or *meteorological instrumentation.* Total stations and electronic distance measurement, as well as other land measurement instruments, are often called *surveying instrumentation.* These may be coupled to laser technologies that are capable of movement in both horizontal and vertical planes, calculating geometry, and may also interface GPS equipment. The measurement of aerosols in the air may couple LIDAR and other sensors relative to road networks or other sources of airborne emission, using portable gas chromatography instruments and mass spectrometry.

Instrumentation may be employed for indoor or outdoor use. Some instrumentation may be used for large-scale data collection and applications, while others may be used for small-scale applications covering very large distances. There are numerous types of sensors and instruments that may be employed for the purposes of gathering spatial information. Each of these usually is digitally interfaced rather than analog interfaced, although many analog signals may be converted to digital format using analog-to-digital converters. There are literally thousands of possible uses for instrumentation:

- *Wireless:* smart phones with GPS capabilities, pagers with GPS capabilities, E911/SOS phones, child and personal locators, asset trackers
- *PC-based products:* autoPC/in-car computers, portable PCs with GPS capabilities, PDAs with GPS capabilities, GPS/GIS software
- *Automobile products:* car navigation systems, car security systems, auto emergency response systems, telematics systems, fleet tracking systems, data loggers
- *General consumer and marine products:* recreational and entertainment GPS products, wristwatches with GPS capabilities, portable navigation systems, marine handheld GPS systems, pet locators

Sensors and instrumentation are designed to monitor and measure the nature of phenomena—to grade them. In a geotechnology sense, this includes more specifically the determination of attributes and their values over time. When coupled to sensors, GPSs provide location and navigation and can be used to guide a sensor through a series of waypoints, automatically measuring events and identifying their positions. GIS is then used to manage, analyze, model, and represent this information. As Tansley said in 1914: "The mere taking of an instrument in the field and the recording of observations...is no guarantee of scientific results" (Tansley, 1914). Knowledge about the phenomena and the variables to be measured may take either as a whole-area approach or a point-based approach (Burrough, 1987). The *whole-area approach* results in the measured variables being transformed to larger areas, while the *point-based approach* involves the sampling of a variable and interpolating the variable.

Both approaches can be used where sensors and instrumentation are involved. In the simplest form a whole-area approach could be the collection of air temperatures for a large area, averaging them all together and applying the average to an area—a map of regional average air temperature would be an example. Using a point approach, air temperatures are measured and interpolated, providing scaled values across the same region. As one moves from a point-based to a whole-area approach, we begin to speak about modeling and simulation—the need for representing phenomena and variables which are not directly measured. However, our interest for the moment is to acquire information for variables using integrated geotechnologies interfaced to sensors and instruments. In a geographic information context this might be referred to as "the collection of spatial and temporal information about variables in either a continuous or non-continuous manner using digital sensors and instrumentation for the purposes of spatial and temporal data acquisition, monitoring and assessment."

Instrumentation and sensors are readily coupled to geotechnologies and interface computing systems. For some sensors and instruments there is specialized software that can process the data. Alternatively, data loggers capable of collecting information from sensors may perform processing and scaling of values prior to output. Data logger output may remain onboard within the data logger, or data may be transferred to data storage models or via telecommunications link.

7.2 LASER OFFSETS AND GPS

Laser offsetting is a technique used to collect GPS waypoints that may not be be collectible normally, due to the inability to track satellites. In such cases the GPS signal is obstructed or blocked by natural conditions or human-made structures. There are times when a GPS could track satellites and gather a waypoint, but to actually stand on the location is neither possible nor desirable. In both cases, waypoint positions cannot be accessed, for a number of reasons:

- Beneath forest canopies
- Under bridges
- Contaminated environments
- Located in bog or wet areas
- Located in traffic
- Located high on a mountain or low in a valley
- Need to avoid entering a research zone
- Military reasons

GPS does not work particularly well under denser forest canopies requiring other considerations and approaches. The foliage blocks the GPS signals and may also result in multipath errors due to signals bouncing from the foliage. Wet foliage is also known to attenuate GPS signals. Consequently, GPS work when the forest is dry is preferred. Some environments (i.e., rainforest) do not provide for these conditions. In deciduous forest conditions, users have located waypoints using GPS by performing working when the foliage is off the trees during the winter months. This is possible provided that the forest is deciduous or has a large composition of deciduous trees. In coniferous stands this may not be the case. But it also depends on the purpose for which the GPS waypoints for are being collected.

As an example, if GPS waypoints are being collected for physiological growth, the measurements are obviously performed more properly in the nonwinter months. Let's first look at a coniferous forest stand. These stands may be unevenly aged, having both young and older trees, or may be plantations of evenly aged trees. Our objective is to collect GPS waypoints at various points within the stand using laser instrumentation coupled to any existing GPS waypoints whose coordinates have been located. To do this we assume that five positions within the forest area need to be sampled. These could be points used for outlining future cut blocks, or they could simply be areas that have incidents of disease for which waypoints are needed. Five sample areas need to be located accurately (Figure 7.1).

The known GPS consists of the square black box in the center of the image. It is surrounded by lakes and located on a road leading to other roads. Five black dots represent each of the waypoints for which positions are needed. Keep in mind that each dot is under forest cover. There are a few ways to approach this problem. One way is to measure the angles from the known point, then measure the distance to each of the five waypoints. Using the angles and distances, each waypoint position can be calculated and tied into the known point. For this method all that is needed is a compass and forester's chain. How accurately can one be with a forester's chain? Most foresters can be pretty accurate, within a few meters or so. Nonforesters may not be as con-

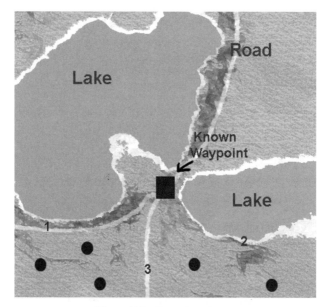

Figure 7.1   Forest locations.

cerned about accuracy; therefore, they could use a laser rangefinder. A laser range-finder is a small binocular laser that is used to measure distances. They are not very expensive and have accuracies of a few centimeters over a few thousand meters, par-ticularly if reflectors are placed on the distant objects. There is one issue that must be considered with a laser rangefinder—they cannot see through trees. Given that these are densely populated forest stands, the user may only be able to proceed using small distances before reaching the waypoints. Therefore, they must keep track of their measurements, tallying them all once the waypoint is reached. For the laser range-finder method, a compass is also used to ensure that steady direction is maintained.

Since points 1 to 3 are on roadways, new waypoints may be determined for each of them. Working from point 1 using the method described above, three points could be accessed. Two points could be accessed from point 2 by working from a known position along the lake edge where GPS measurements would probably be easy to ob-tain. If point 3 were located, three waypoints could be located from its road position to the left and two points to the right. From point 3 all the points are closer to the dots and would decrease possibilities of compass and laser errors due to the shorter dis-tances. The principle behind offsetting is to get as close to the desired waypoints as possible, then to measure from these georeferenced positions to unknown positions while determining the angles. In North America, where many forests are in their nat-ural state, there are bound to be openings in the forest canopy. These provide places where known GPS waypoints may be determined and from which offsets can be cal-culated. In Europe, where there are more plantations, forests tend to be even aged, regularly spaced, and therefore lacking such openings. However, plantations are also not continuous in most cases and roads are closer to their periphery. The concept of using GPSs, lasers, and compasses together for offsetting applies not only to forests; it can be used within urban jungles, under bridges, and even inside buildings. Once the distances and angles are determined, the known positions can be offset based on that information to create new points.

Finally, one of the issues with this method and integration of technologies is that the manually collected information from traversing will need to be entered manually into the data set along GPS readings from known positions. However, there are some laser rangefinders that can store the data along with GPS readings. Total stations are good examples of this type of equipment. The problem for foresters in using a total station is the scale and ruggedness in the forest. A total station needs to be set up prop-erly before acquiring accurate information. That may be fine for urban environments but very impractical where tens of kilometers of rugged terrain must be traversed, often in poor weather precluding continual setup.

7.3 DATA LOGGERS

A data logger is usually a small portable processing unit that is able to integrate ana-log and digital signals from sensors (Figure 7.2). Data loggers may be very inexpen-sive and contain inputs for one or two sensors. Others may be very expensive and ca-pable of interfacing numerous types of sensors in varying quantities. In application a

Figure 7.2   Field data logger.

data logger is maintained on a site, powered internally by batteries, direct line, or solar panels. The choice of power supply depends on where the site is. If the site is close to the office and good roads allow for access, the data logger may have a battery or direct power supply. But if the site is in a remote environment, a combination of batteries and solar panels may be utilized.

Thus data loggers can be found in all types of environments and are usually tested to withstand harsh environmental conditions and provided with appropriate enclosures. Sensors are attached to a data logger via input channels. Each channel is numbered and programmed according to the sensor type and sampling frequency desired. The data logger scales the input signals from each sensor, recording the values and assigning the results according to the time. As an example, a soil temperature sensor may be inserted into the ground and the soil temperature monitored and recorded within ±1.0°C for each 10-minute period. Alternatively, six values for 1 hour could be averaged together to come up with one average temperature for the hour for that one sensor. Each sensor may be averaged or groups of sensors may be averaged. Depending on the data logger, maximum, minimum, and standard deviations may be determined for an individual sensor or for groups of sensors. Other statistical information is possible in the more expensive types of data loggers. To provide processing of the scaled values, some type of onboard intermediate processing capability is needed.

Data collected in the data logger can be stored internally within the data logger. Internal storage is used where data loggers are located in remote northern areas or other locations where real-time connection is not needed. Applications based on road condition sensors involving data loggers require direct links for constant monitoring of conditions. In these cases, telephone dial-ups are used and data transferred from the data logger to office computers using appropriate software. In a remote environ-

Figure 7.3   Data logger with meteorological sensors.

ment the data logger must have sufficient storage capacity to record information from the sensors, sometimes for months.

Storage is usually in the form of solid-state storage and comes in a modular form of some kind. If high sampling intervals are used, very large amounts of data may be recorded, especially where numerous sensors are attached to the data logger (Figure 7.3). It is not uncommon to find data loggers with 20 or more sensors attached to them, recording information hourly. In this case, 480 values are being recorded per day or about 14,400 values per month. This does not include other output, such as maxima–minima or averaging. For some applications that are not visited for months, onboard data storage requirements may be large.

7.4 SENSORS

The selection and installation of sensors are related to the entity that is to be measured and its characteristics. Temperature sensors can have varying resolution (±0.5, ±1.0, or even coarser, $\pm2.0°C$). If the entity to be monitored and measured does not vary in temperature much more than $\pm1.0°C$ over time as an example, it would be pointless to use a sensor that cannot represent the subtle changes in temperature present. In this case even a sensor with $\pm0.5°C$ accuracy will cover a $1°$ temperature change. Therefore, a sensor capable of finer resolutions would be desirable. Sensor accuracy and resolution also apply to other sensors. Not all sensors are linear; most have an optimum range where their signal characteristics are more linear than others. A temperature sensor may function from -20.0 to $+50.0°C$ but is most linear when above zero but below $+35.0°C$.

Rainfall is often measured using a tipping bucket rain gauge. In operation, rain falls into the gauge, filling a small bucket, which then tips when a precalibrated amount fills the bucket. The tip of the bucket usually causes a magnetic switch to register a pulse. Each pulse is counted and represents one event or tip for the calibrated amount. This sounds simple enough; however, there are many anomalies to consider about a rain gauge. These include the storm itself as well as the calibration of the rain gauge. Some rain gauges may tip with only 0.2 mm in the bucket, as compared to some gauges that tip when only 1.0 mm of rainfall has been collected in the bucket. If the storm does not include heavy rainfall periods, and they are intermittent, it is quite possible that 1 mm may not be accumulated to cause a tip. In such cases the rainfall evaporates from the tipping bucket before the next storm period. Alternatively, a storm of high intensity may approach with heavy downpours, which fill or overflow the rain gauge. Since the tipping mechanism can accept rainfall only at a certain rate, it cannot register the true rainfall event (rate).

Snow depth is difficult to measure. First, an instrument is needed that can operate in subzero temperatures. Traditionally, snow depth has been measured along predetermined routes. These sampling points can also be located using GPS. A snow ruler would be inserted into the snowpack and the depth measured and recorded manually. There is a great advantage to measuring snow using a route of waypoints. The snow depth measured at numerous points is more representative of the variation over the landscape as the depth measured for a single point. A disadvantage of this method for sampling snow depth is that it is disturbs the locations. Footprints during access cause large depressions that lead to later drifting and accumulation, which then falsely represent the snow depth.

Alternatively, snow depth sensors are used. Connected to a data logger, snow depth sensors usually operate by sound or optical beam. In the case of sound, high-pitched ultrasonic waves are bounced from the snowpack back to the snow sensor. Because sound travels at different speeds depending on temperature, the air temperature is usually included in the algorithm and the programming for snow depth sensors. For optical sensors a light beam is emitted from the sensor to the snow surface and reflected back to the sensor. The return time of the signal is a measure of snow depth. In practice, snow depth sensors are usually located in open areas (although not always) and remain permanent, similar to other climatic sensors. This applies to long-term weather stations where climatic records are kept over time and can readily be compared based on one location. But what if a nonpermanent study is required, lasting perhaps only one winter or more? For these cases it is not unusual to find a series of snow depth sensors distributed over a geographic area. This information is then exported to GIS and can be analyzed both spatially and temporally.

Snow density is often the more valuable measurement—and the more difficult to acquire. Whereas snow depth provides the level of the snowpack, snow density provides an indication of the water equivalent within the snowpack. Snowpack density may vary considerably over very short distances. Subsequently, it is difficult to measure this variation accurately. The snow density also varies by depth. A snow depth sensor measures the height of the snowpack above the bare ground level, not below the snow surface. To measure the snow density, cores can be taken from selected GPS waypoints. The cores can be divided into a series of depths and the water equivalents determined by depth by melting the individual cores. In this way, water equivalent is calculated based on the volume and amount of water for individual segments of each core. Pressure transducers have been used to measure snow depth. A pressure transducer records varying pressure at the sensor. A series of pressure transducers inserted into a snowpack profile provide varying readings throughout the snowpack. In operation a snowpack may have five or more pressure transducers inserted in a vertical fashion. Since time is also output, this information can be exported to GIS and represented in 3D. Color ramps can be used to show the varying snow density using either cores or pressure transducer measurement methods. Contours can also be generated using GIS for snow density for either a horizontal or a vertical representation.

7.5 DATA LOGGERS AND VISUALIZATION

Spatial and temporal information collected with sensors is creating new opportunities in the area of visualization. Since the data sets are continuous, they can be developed into effective visualizations using either animation or immersive environments. For immersive environments the data may also be explored using data mining techniques. Individual measurements may be represented using symbols (e.g., circles, squares, triangles) for individual or groups of sensors that are distributed in both time and space. It is not possible to review data logger files for large multisensor applications in a quick, efficient way using traditional techniques, unless significant elimination and classification of values is conducted first. In an immersive environment, data values can be explored and "walked through" as they appear in space. Using this technique, relationships between values can quickly be observed prior to further investigation and analysis of the data set. A key value to using immersive techniques for exploring these data sets is that trends are readily observed and outliers are identified. The viewer can formulate hypotheses in the immersive environment.

Using a 1-minute sample interval could conceivably result in 1440 thematic layers for a sensor on any given day. Collectively, these thematic layers could be used to form animations that are very smooth in transition, although they would also be very large in size. Continuous values are useful for constructing near-realistic visualizations and can emulate processes very well. There may, however, be considerations about how many data can be represented at once from an array of sensors.

Imagine that there are 40 sensors with output occurring every 5 minutes. How can those values be represented on a display screen in the form of a map? How can the changing temperature, light and wind, and plant growth be represented and viewed all at once? Obviously, it would take a fair amount of creativity and thought to represent this information in an integrated manner. Multirepresentation using split screens and/or dual display monitors could prove useful for this purpose. Ideally, these visualizations would incorporate as many data as possible and allow for one to view the interrelationships between processes and phenomena.

The coupling of GIS to visualization is resulting in new possibilities for the exploration and communication of these types of data. In the forestry area, photo-realistic

Figure 7.4 Sensor data and visualization.

vegetation is being used regularly in visualization. The vegetation is placed on a DEM along with other 3D objects (Figure 7.4). This applies to urban visualization applications as well. The forest inventory becomes the source of data for populating the visualization. Going further, sensor data can be represented on the landscape through the use of color camps and symbols. These visualizations then incorporate thematic layers from both forest inventory and sensor data, providing a means to visualize the sensor information as it relates to objects. This is a very useful and powerful way of communicating these types of relationships to others where themes can be used to "tint" landscapes.

Since visualization software can link to GIS databases, any combination of sensor data and objects can be rendered. The scale for these representations is probably large scale since sensors are in practice applied only to small areas. They might be represented in small scale, however, where groups of sensors or multiplexed sensors are being used in a network manner. For large-scale representation, consideration has to be given to the 3D objects used along with the sensor information. Detail becomes more critical in these smaller areas (< 500 m^2).

7.6 SENSOR PORTABILITY

Some uses of sensors require that the sensors be mobile; these include emergency applications, environmental applications, and military applications where data are to be gathered over large areas, precluding small-area representation. In some of these cases the environments may not be suitable for human access. Areas with high radiation areas, toxic chemicals, and armed threat are examples of these environments. Others include locations where it is simply impossible to access the positions required. Robotics, kites, large and model aircraft, balloons, and automated transportation equipment may be used. Sensors and instrumentation are attached to these devices, and they move around a location with onboard data loggers and sensors collecting information.

Let's assume an imaginary application that involves numerous geotechnologies. Buildings exist for a three-square-block area. From the CAD or GIS architecture drawings, 3D objects of the buildings can be constructed. A DEM exists and shows that the area is on a hill with the slope oriented toward the prevailing wind. The buildings can be located in GIS and situated with respect to DEM coordinates. Visualizations are constructed of all this information for the purposes of communicating and exploration. One of the concerns expressed is that the buildings result in a tunneling of wind between them. Someone has considered that this effect may, in the event of fire, result in the fire spreading quickly from building to building. In GIS, building are analyzed with respect to their location and height. Shade profiles between buildings can be ascertained that show both lines of site and how one building shades another. As no one knows for sure about the wind between the buildings, the need exists to acquire some information about the wind. How can that information be gathered? Can it be modeled? If so, from what data currently existing? To do this, it was first decided that several wind speed sensors (anemometers) be located on the ground between the buildings. From that initial data set it was found that the wind speeds varied significantly (x,y) for a series of coordinates (waypoints). The question arose: Does the wind speed vary by height? It probably does. Therefore, after careful consideration a few methods were investigated for collecting the wind speeds between buildings.

The first method involved hanging sensors from the sides of buildings throughout the area. This had advantages because they could be deployed quickly and wired through buildings, making data collection easier. The drawback to this method is that it did not measure the winds between the buildings but instead, the wind at building surfaces. Keeping in mind that the overall objective was not only to understand the nature of wind between buildings but also to develop a strategy for reducing them, it was decided that more information about wind speeds between the buildings was needed.

To acquire this information, small balloons were first used with sensors attached to them. These were deployed between buildings and their positions determined using laser and GPS offsetting. The sensors recorded the wind speeds for 10 days, the data were analyzed, and a 3D representation of wind speed was constructed for the entire area. Although it could be expected that major storms would appear from time to time and that higher speeds may be possible, these data provided a means to begin modeling the wind speeds between buildings. After some time a strategy was developed to reduce the wind speed between buildings which involved the planting of trees throughout the neighborhood to act as barriers to reduce the wind. This also had positive benefits in terms of aesthetics.

But the application did not stop there. With knowledge of the potential winds, thus potential rate of spread, other ideas about the spread of aerosols were investigated. If there were a gas leak or some other aerosol release into the atmosphere, the wind speed information could be modeled to show where it could be expected to travel and at what speed. Finally, an analysis of pedestrian traffic patterns and vehicle movements provided a means to determine emergency access into the area as well as providing optimum routes for leaving the area in the event of an emergency. What began as a study of the wind resulted in the development of new applications. This is common where new data and new geotechnologies are applied.

Other methods for collecting data using sensors and instrumentations involve the use of boats, bicycles, quads, and hang gliders. Quads are four-wheeled motorcycles that can move over rough terrain very quickly without becoming stuck in bogs and mud. They are small enough that they are useful for moving through narrow streets, and they can also carry a fairly large amount of weight, due to the floatation created from large tires. In many environmental applications quads are used for moving across the landscape.

GPS is used for navigation in remote areas, and a data logger is strapped to the quad. Sensors are connected to the data logger and the rider drives from sample location to sample location and takes measurements. The sensor information is integrated with the GPS location and there are digital time stamps for both sets of data. A similar scenario exists for boats: A GPS can navigate a boat to selected locations on a lake or body of water, with depth sensors or water quality sensors used at those locations. The data are once again integrated and recorded for later analysis.

Sometimes a shoreline and other open areas can be mapped from one position through the use of integrated technologies. Choosing a central point with known position as determined by GPS, laser measurements are made to shorelines. Since both distance and angle can be determined, shoreline coordinates can also be determined from these measurements. This technique also works in open valleys, high locations, and other areas where boundary edges need to be mapped and is termed a *radiating survey*—all points radiate from one location. This is an effective technique where the waterline changes due to drainage and recharging in watersheds. Lake levels may not change appreciably from an aerial photograph perspective or remotely sensed image because of resolution. A 20-cm rise in lake level may not be detectable on a raster-by-raster basis but can be measured from a central location using radiation survey techniques. Why is this important? Because although undetectable, a 20-cm rise in lake level for a very large lake may represent a very large change in the size of the lake, perhaps hundreds of hectares or more.

7.7 GPS AND TELEMETRY TRACKING OF ANIMALS

Several organizations are interested in tracking the movements of animals across the landscape. This provides information about animal behavior as well as the habitats associated with them. They are used to monitor fish movements, moose, caribou, elk, deer, bear, and other wildlife and have been used most often in oceans, forests, and national parks (Thurston and Pruss, 2000). Traditionally, animals have been tracked using radio telemetry. Radio telemetry involves fitting on an animal a small radio receiver that emits a signal. That signal is then ascertained using radio telemetry receivers. To acquire the position of the wildlife being tracked, the telemetry receiver recovers a number of signals from the animal. The location of the animal is then determined through the process of triangulation, narrowing the location of the animal to a smaller polygon area. The accuracy of radio triangulation depends on several factors:

- Whether or not the animal is moving during the time being tracked
- Rate of animal movement
- Time of day as it relates to whether or not wildlife is likely to be moving
- Signal strength of the telemetry equipment
- Error sources generated by nearby surfaces (i.e., multipath)

Once several locations (polygons) have been acquired for wildlife positions, these polygons can be georeferenced. The centers of the triangulated areas are established as points and the data are imported into GIS for analysis. Since a number of triangulations for an individual animal can be made rapidly, the bounding area or extent for any given triangulated polygon area may be larger or smaller—demonstrate higher or lower accuracy. In general, using radio telemetry equipment, wildlife can be tracked reasonably accurately to within an area 1 to 4 hectares in size.

In operation, this process may occur over the course of the day, week, month, or even years. Each successive set of locations can then be tabulated into database tables, forming a record of wildlife locations related to date and time. Using a GIS, spatial analysis can be performed using the data set, querying independent or grouped variables for all wildlife locations. This is not an easy process. It requires considerable effort in labor and field visitation, which becomes more difficult depending on the regularity of field visitation. Many animal tracking projects begin with several animals fitted with radio telemetry equipment for the purpose of ensuring a representative population that can be analyzed statistically at a later date. In some cases, for longer-running investigations, wildlife mortality due to natural or human causes can reduce the sample population. This presents a problem when attempting to acquire a sufficiently large data set that can be interpreted scientifically.

The animal locations are imported into GIS as a layer, together with topographic data sets or DEM and often with vegetation layers forming another variable. Through the study of animal locations and topography, animal movement in relation to topography can be determined. Similarly, analysis of vegetation layers together with animal location can provide a means of understanding animal habitat preferences. This same concept can be applied to humans by studying their movements through communities, stores, and recreational areas.

Radio telemetry is relatively quick and easy to use. Although its accuracy is less than that of GPS equipment, it should be remembered that many vegetation maps are generated from Landsat imagery with a 30-m pixel resolution. Thus, the vegetative layer will have less resolution than the accuracy associated with radio telemetry equipment. In the case of DEM, they are often generated using 100-m-spaced grid samples, again resulting in lower levels of resolution. Studies involving radio telemetry are therefore well suited for use of vegetation and topographic information gathered at these resolutions. The advantage of using radio telemetry is reduced cost for equipment and the purchase of related products. The disadvantage is that animal locations can only be tracked while the user is in the field—wildlife are not tracked continuously. For continuous tracking of animal movements, alternative technologies must be utilized.

GPS-based animal tracking collars and devices have become more popular due to the fact that they provide continuous tracking capability, lower field costs, and higher accuracy. Small GPS devices are fitted to wildlife that monitor wildlife location continuously using GPS. These devices can be preset for variable sampling intervals, usually hourly, but depending on the nature of the wildlife being tracked, this may be increased or decreased significantly.

The rate of sampling is a critical factor when initiating GPS-based tracking systems. Although there is greater value in having shorter times between location fixes, this information must be stored in the device. Sampling too frequently can fill the storage memory of the GPS device, thereby shortening the duration of data collection. This becomes even more of a problem if the locations being recorded are a function of time only, as compared to quality. In GPS terms, positions having (x,y,z) locations are the most useful. Therefore, tracking devices with intermediate recognition of location quality, which record only the adequate positions, are much more useful and can be used for higher sampling intervals.

Conversely, lowering the sampling interval for remote GPS-based tracking devices can result in longer tracking durations. The downside to this is that some wildlife movements may not be recorded, as they occur between longer sampling intervals. This would be analogous to having a chart showing point A and point B but not showing the points in between, which might be the more useful information. These are important considerations in remote environments where GPS tracking devices are fitted to wildlife that are self-monitoring and being interrogated periodically. In such cases the user visits the field periodically, downloading the data from the animal. Needless to say, the animal must be found and confronted and dealt with to download the information. That itself may be a disadvantage.

Real-time animal monitoring involves fitting to wildlife a GPS tracking device that can be interrogated continuously. This means that the tracking device must also have a means to telecommunicate the GPS information from the animal to a GPS platform that is capable of connecting to and downloading the GPS information from the collar. The advantage of this approach is that a high sampling frequency can be used because it does not depend on remote storage. Due to that fact that GPS equipment (uncorrected) has a published accuracy of ± 37 m ($x-y$ axis), this can result in very detailed location information. Many of these applications using the real-time approach often provide information not been previously attainable that sheds new light on the nature of wildlife. GPS information is also highly compatible with 30-m-resolution satellite imagery. Since GPS often achieves much higher resolutions than the published accuracy (i.e., <15 uncorrected), aerial photography, with its higher resolutions, may become more practical to use. The advantage of aerial photography is that higher-resolution DEM can be constructed, and photo interpretation may take place, providing other information useful for the investigation at hand.

7.8   SENSORS AND MODELING

There are two broad ranges of models, structural models and relational models. Structural models include object models and action models. Object models are static and represented from diagrams, whereas action models represent moving entities (Poiker, 1999). Models are used to represent real phenomenon and are constructed from real data. They vary in purpose, scale, and extent. Some model very small areas, whereas others, such as global models, seek to model processes occurring on a planetary scale.

There are other types of models, including classification models, spatial analysis models, distance models, and distance and neighborhood models. Location–allocation models are used to determine placements of stores and for transportation purposes. They consider population demographics and spatial distances. Consumer preferences vary and information about consumer movement can be collected with optical sensors and photographic recording instrumentation. Gravity models are based on distances between features—the closer they are, the greater the likelihood that they influence each other. Measuring these influences can involve sensors and instrumentation. This is particularly true where biological phenomena are being measured with respect to varying distance (Murphy, 1996). Ecological models are developed from the measurement of many agents within biological systems. Where changing data over time are incorporated into a model, we refer to a *simulation*. Simulations include disease spread, plant growth, animal movement, population increases and decreases, hydrological processes, soil interactions, flooding, hurricanes, and population movements.

Atmospheric models have become very popular in recent times, due to growing interest in climate change. These models are related to human activity and biological processes to determine their contributions. Sea-ice volumes are also affected by climate warming (Govindasamy and Caldeira, 2001). The wind power sector grew by 35 percent in Europe during 2001. Wind is monitored, analyzed, and distribution determined using many geotechnologies coupled to GIS. Hourly emissions and flow rates, as well as seasonal partition of data, is being used to model air quality (Houyoux, 1998). There are numerous applications, perhaps tens of thousands internationally, that involve the use of sensor and instrumentation technologies for measuring agents.

Useful models are developed by accurately defining the indicators that represent the processes of the model being considered. The construction or choice of indicators is a major topic in the sustainable development discourse. Operationalizing and quantifying sustainability is about nothing less than the attempt to grasp, and finally manage, passing from today's world to a future sustainable world state, but the debate on the right type of measurement is far from being closed (Scheller, 2000). Although many basic physiological relationships between climate change and plant growth have been addressed since the inception of the U.S. national global change research program in the early 1990s, understanding the interactions between stresses on individual trees, especially on multiple stress interactions at the forest level, are still very limited (Aber et al., 2001).

A common theme in current climate, ecosystem, and global change modeling is the need for more information. To acquire this information, sensors and instrumentation will need to be used on coupled to geotechnologies and analyzed using GIS. This information will be acquired on both large and small scales. There are many opportunities in this field for geotechnology professionals.

7.9   RISK, THREAT, AND SECURITY

Area security is in essence an elaborate exercise in risk management. Governmental and commercial leaders will never have the resources necessary concurrently to create and defend geographically dispersed strong points, although natural defensive thought predisposes this course of action. Moreover, how can crisis management tendencies be avoided so that an overzealous security reaction does not excessively restrict the essential activity it intends to protect? What mix of civil unrest and criminal activity can be tolerated before the critical capacity is threatened? What is the measure of threat? What methods can be employed to enable the economic increase in security measures and forces that are prudent and acceptable to cost-wary executives, stockholders, or taxpayers? How do joint, interagency, contract security forces, and emergency services synchronize their operations for mutual support? Typically, law enforcement special tactical elements are limited in the number of concurrent incidents they can respond to. Additionally, they are typically one component of a larger crisis/emergency management effort comprised of emergency medical, fire, and bomb-disposal personnel that requires additional public safety augmentation for cordon, search, and evacuation operations. In the case of a natural disaster, add utility crews and public shelter/relief to the choreography. Confusion will be a factor that increases with each echelon and agency seeking to help resolve the situation.

Vulnerability assessments and security reviews often result in recommendations that require major funding. However, operational art and value added are demonstrated in the suggestion of interim adjustments to procedures that return immediate security enhancement and buy the time required for system and staffing upgrades to receive approval and funding. The most significant and immediate enhancement to security and consequence management can be accomplished through visualization support to decision makers by using GIS as a cornerstone in a multidisciplined approach to planning, rehearsal, and execution during crisis situations (Figure 7.5).

Current GIS tools make it possible to transform the multitude of disparate data sets developed and used routinely for daily management, maintenance, and accountability from lists and spreadsheets into interactive and intelligent map-based visuals. This spatial visualization makes obvious the significance of otherwise esoteric information. Mapping and comparing site functional patterns with organized compartmentation and access controls may identify where procedural modifications can reduce vulnerability without restricting operations.

Mapping the physical layout and distribution of permanent and temporary objects and their effect on observation from outside, as well as security forces, observation, and fields of fire, may suggest simple cleanup that will passively enhance security on a specific site at a micro level but applies to manipulation of municipal/regional traf-

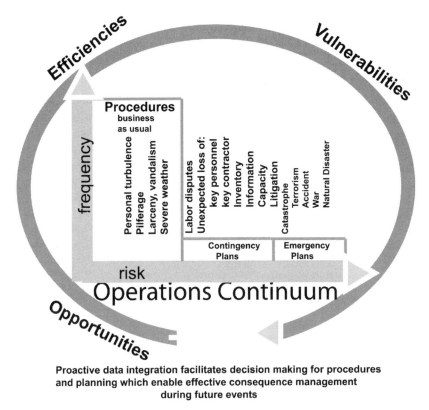

**Proactive data integration facilitates decision making for procedures
and planning which enable effective consequence management
during future events**

Figure 7.5   Risk, threat, and security operations continuum (see the chapter appendix).

fic patterns at a macro level. Identifying choke points for material and information handling that define the most critical aspect of a site or activity will establish a natural priority for security measures designed to restrict or at least monitor collateral activity in and around these locations. The presence and location of volatile materials especially susceptible to divisive exploitation and hazardous-materials potentials must be appreciated, as well as responders within the situational environment as it develops—the presence of utility services such as high-voltage electrical; steam; heating, ventilating, and air conditioning; and reaction to heat and water. Additionally, public safety, emergency services, and security forces must know utility service vulnerability points and redundancies, as well as the terrain surrounding the site and what advantage it affords the threat. Threat scenarios might range from external protest/ picketing to penetration activities (e.g., vandalism, sabotage, occupation, hostage taking) to actual attack. Finally, the procedures should be checked with integral after-action reviews (AARs). In the absence of unlimited budgets, imagination and initiative are the keys to success. Appreciation and optimization of the existing datascape through GIS visualization will pay dividends for decision making during crisis situations by enabling a smooth transition from more effective routine management.

For security planning to be successful in areas of mixed jurisdiction, course of action (COA) development and war gaming must include civil authorities, industry, and interagency participation using the actual jurisdictional context as a framework for developing roles and responsibilities that are appropriate and legal. Operations planners can then develop the most practical interagency task organization, which may necessitate revisions in the roles, responsibilities, and relationships of all security forces and emergency services. Although state and municipal emergency management divisions (EMDs) routinely tested the incident command system (ICS) prior to the terrorist attack on New York on September 11, 2001, the majority of scenarios centered on consequence management resulting from natural disasters. Typical scenarios focused on a single natural event with minor opportunistic criminal activity. However, a determined and protracted terrorist threat creates conditions that take advantage of the tendency to commit all available resources to an incident immediately.

Although the initial task would be to conduct a technical assessment of vulnerability relative to existing security plans and procedures, a description of operational mechanics and a catalog of points of interest will not satisfy the principles necessary to account for vulnerability or provide operational assurance. The task is scalable but can be intimidating because of intricacies in community relations and ambiguity in the situation that determines jurisdictional authority and responsibility. Appropriately, current events have heightened concerns for the physical security of societies.

Although much of the current focus is naturally on physical threats, it is important to acknowledge that virtual attacks manifest physical consequences. Incorporating this dimension of systemic interdependence/operability can only be achieved through the visualization provided by GIS. Decision support must enable fusion of epidemiological and public health information for indications of a chemical/biological attack that is best understood and tracked by the appropriate agencies. Sensor technologies coupled to geotechnologies provide a means for capturing some of these types of information. For cost-conscience municipalities and commercial activities, optimal preparedness is achieved by creating opportunities and gaining efficiencies within routine daily processes that provide indications of vulnerability and threat so that risk can be avoided and consequences mitigated.

7.10   MEDICAL GEOGRAPHY AND EPIDEMIOLOGY

Medical geography can be defined as the branch of geography concerned with the geographical aspects of health (status) and health care (systems). It seeks to broaden our understanding of the various factors that affect the health of populations (Anon., 1998). The suggestion that geographical location has an influence on the status of health is an old concept in Western medicine. Hippocrates (460–377 B.C.), the father of modern medicine, observed that certain diseases seem to occur in some places and not in others. Medicine entered the age of reason based on observation. Hippocrates believed that illness had a physical and rational explanation and studied such factors as climate, water, and diet and their effects on disease. The more information gained about why certain diseases seem to occur only in certain places and not in others has frequently led to new insights into the nature of disease itself.

The best-known example of mapping disease to gain greater insight into its origin took place in September 1854. During an outbreak of cholera in London, John Snow, a local physician, decided to test his hypothesis that cholera is a waterborne disease and that the outbreak was a result of contaminated water supplies. This view was contrary to the medical beliefs of the time. By drawing a map showing where the victims lived and where the local water pumps were located, Snow was able to show a clear clustering of cholera cases around one of the water pumps, which was drawing water from a known polluted section of the Thames river. Based on the evidence depicted on his map (Figure 7.6), Snow shut down the pump and the outbreak of cholera subsided. By mapping the information, Snow could easily pinpoint the instances of disease vis-à-vis the location of water pumps and thus demonstrate a cause-and-effect relationship between the water and the disease (Cliff and Haggett, 1988).

Although Snow's examination of the cholera epidemic was significant, analysis of the spatial distribution of diseases has been fairly limited in the medical field, due to

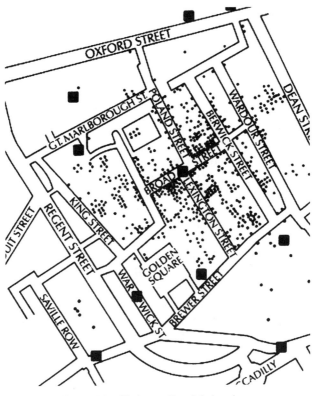

Figure 7.6   Cholera outbreak in London.

the complex and time-consuming nature of working with maps. The technological revolution of recent years has provided powerful processing platforms that have helped in the development of computerized spatial mapping systems. Current GISs provide capability and functionality far beyond the manual methods employed by Snow. Ten to 15 years ago, GIS analysis was limited to workstation UNIX computers and command line software packages, but today, user-friendly Windows-based software and inexpensive personal computers have put high-end GIS tools in the hands of various disciplines, including the medical professionals.

Geographic information systems allow medical geographers to collect and analyze data far more easily than was possible with traditional analytical methods. The different types of information in a GIS are used like a set of clear transparencies laid one on top of the other, allowing analysts to understand the relationships among the various data layers, as Snow did with cholera.

Epidemiology is concerned with identifying patterns in the distribution and frequency of a disease, and by using these patterns attempts to identify the causes of disease and to develop appropriate strategies to control or eradicate disease. Medical researchers have long been examining the distribution of disease and death in an attempt to determine if factors of the social or physical environment play a role in the incidence of a disease. Infectious diseases add a temporal dimension to the investigation, that of distribution of the disease through space over time. Epidemiological principles of determining the distribution and cause of a disease provide a foundation for the analysis of a disease incident. These principles need to be understood by GIS analysts in order to test hypotheses about cause-and-effect relationships and to evaluate data quality, confounding factors, and bias that may influence the interpretation of results. Epidemiologists have traditionally used maps when analyzing associations among location, environment, and disease (Gesler, 1986), and we now need to be able to understand and evaluate maps prepared using GIS technology.

A GIS also provides an epidemiologist with much greater potential for developing clear hypotheses or results through its display, spatial, and statistical analysis capabilities. These allow the user to examine and observe medical data in new and highly effective ways. Spatial analysis is the application of analytical techniques associated with the study of locations of geographic phenomena together with their spatial dimensions and associated attributes. It is useful for evaluating suitability, for estimating and predicting, and for interpreting and understanding the location and distribution of geographic features and phenomena (Walker, 1993). Spatial analysis encompasses many methods and procedures developed in geography, statistics, and other disciplines for analyzing and relating spatial information. Spatial relationships based on proximity and relative location of entities form the core of spatial analysis, and its complexity can range from simple map overlay to statistical modeling. Visualization is also an important capability of a GIS and has been used to show a change in disease patterns over time. Animation is also highly effective in depicting the spread or retreat of a disease over space and time (Clarke et al., 1996).

Although relatively new in the epidemiological field, the development and use of GIS has gained momentum in other areas as well, such as agriculture, marketing, transportation, law enforcement, municipal planning, disaster response, and environmental monitoring. Epidemiological GISs have recently been used to map the risk of malaria in Brazil, where it was observed that almost half of the cases occur in a single state containing a small fraction of the population, and in Central American and Caribbean countries between 1990 and 1993, where it was also employed to monitor malaria (Pan American Health Organization, 1996). The field of GIS is increasing in importance to medical geographers and researchers around the world for the study of environmental factors and geographical patterns of diseases, and is essential to gaining an advantage or finding the cause of a disease and effective treatment.

A GIS can be used to analyze animal disease outbreaks, allowing a spatial component to be incorporated in the investigation and reporting of an incident. A map is a valuable tool that can provide a visual perspective and additional insight into the research. The mapping of a disease is straightforward; it is the interpretation of disease data that can be the challenge. The extensive data management and analytical and display capabilities of a GIS can provide the tools to make more effective interpretations in research into animal disease. It is often assumed that geotechnology principles are used only in landscape monitoring and measurement; yet on a much larger scale, these technologies are being used to monitor human health and physiology as well as to identify human shape, form, and texture. In addition, a review of the medical literature in the area of image analysis, for example, reveals many of the same issues that landscape managers contend with using geotechnologies, including visualization. These include segmentation of x-rays, MRI, and PET scans; and referencing and occlusion of objects, flow, and volumes in 2D, 3D, and 4D.

EXERCISES

7.1. Data loggers can be coupled to various types of instruments and sensors. If you wanted to build a map of a stream bottom, how would you go about acquiring the information needed to do so?

7.2. Discuss briefly why an understanding of the phenomenon to be measured, then mapped, is important, and how the two relate to each other.

7.3. Describe how a laser rangefinder can be used together with a compass to provide information about snow depths across a landscape.

7.4. Your task is to monitor soil temperatures for five locations, then build a table of spatial and aspatial data suitable for use in a GIS. Briefly describe how you would go about this task.

7.5. What is a data logger, and how can it be used to collect information about solar radiation during the course of a day?

7.6. What concerns and considerations are involved in collecting spatial data using mobile platforms?

7.7. Visualization is useful when exploring and interpreting large data sets. Briefly describe the value of visualizing data from data loggers.

7.8. Can the climate inside buildings be mapped? If so, how would you go about it?

7.9. Cellular-based technologies permit the transfer of maps in real time. Discuss two applications where cellular technologies can be used effectively with maps.

7.10. What is the difference between sampling duration and sampling frequency?

7.11. Assume that an animal has a GPS tracking collar. The objective is to monitor the animal's movements and habitat. How would you go about acquiring the information, and what are the considerations?

7.12. A disease appears to be spreading in a region. Discuss how you would go about mapping the occurrence and communicating the results effectively.

APPENDIX: OPERATIONS CONTINUUM (FIGURE 7.5)

1. All municipal and commercial entities conduct their affairs in a continuum. This operations continuum is punctuated by key events that define experiential transition points for every organization. The distinction of these points is as the culmination to a level of operations. For an organization to grow, it must confront and cope successfully with external influences that it cannot control but which define its operational environment.

2. Generally, the intention and sequence is to seek opportunity, gain efficiencies, and reduce vulnerabilities. The fantasy of leaders everywhere is that their operations will loop back directly to opportunities created through efficiencies with only negligible risk.

3. The key correlation is: Within the intended scope of operations for any organization, the higher the frequency of occurrence, the lower the risk, regardless of the inherent physical danger of the task. Conversely, situations that present the greatest risk are those that are extraordinary to operational routines, regardless of the absence of physical danger.

4. Procedures are established to deal with probable occurrences. The key discriminator is expectation. There are generic challenges that any organization should prepare for. Procedures are intended to enable organic leadership to retain control, establish acceptable efficiencies, and detect and refer threats to the appropriate jurisdictional authority. This category of activity is euphemistically referred to as "business as usual." Additionally, procedures are less affected by personnel turbulence because organizations can easily assimilate new members without loss of effectiveness through mentoring and experience reinforced by routine duties.

5. Plans are established to deal with anticipated but unlikely occurrences. Often, these contingencies are not unique but similar to situations with existing procedures; however, they are associated with severely escalated consequences. The intent is to minimize adverse effects from unexpected situations ranging from temporary operational confusion to an unrepairable loss of capacity. Multiechelon training is essen-

tial. Conceptual communication and physical rehearsals will transform nonstandard intuition and conventional wisdom into common sense.

6. The demarcation between procedures and plans is actual performance. Regardless of how well an organization has rehearsed, all preparations are plans until validated during an actual occurrence, which transforms them into procedures. Once experienced, the organization gains confidence in its ability to develop and conclude a similar situation successfully. This effectively reduces vulnerability by removing operational ambiguity present within the best of plans. This is where organizational maturation occurs. However, the benefits of experience are negated by personnel turbulence, which reduces the qualified experience of the organization, leaving it unaware of actual vulnerabilities by assuming preparedness in that adequate procedures exist, when effectively it should return to the planning process.

7. As an organization is forced beyond its comfort zone of high-frequency, low-risk activity, its foundation of procedures and the quality of its planning will determine recoverability. When situations become catastrophic, emergency management procedures go into effect. During these conditions the objective is survival over prosperity, and a deliberate transition of control to an incident command organization should be anticipated.

8. The emergency management threshold is often the point of departure for discussions of disaster preparedness and consequence management. Although the frequency of occurrence ranges from unlikely to improbable, the risk increases exponentially and has been witnessed in recorded events. Therefore, participation in emergency management activities is accepted as an obligation demonstrating responsible leadership.

9. When taken out of context with the operations continuum, emergency management preparedness is often viewed as requiring specialized staff or hedged resources to guarantee success. This perception causes small commercial entities and economically disadvantaged municipalities to view true preparedness as a luxury they cannot afford. However, successful operations under any conditions depend entirely on the quality of decisions. Ideal control of risk is accomplished through situational awareness (SA) supporting decision making. SA is measured by how completely it considers all aspects of the organization's internal functions and relates them to the influences of the crisis situation. At this time more than any other, the quality of preparedness is determined by the resiliency gained through routine procedures and complementary plans which establish the only foundation providing viable emergency management potential.

10. Routine to extreme effective consequence management starts with "business as usual." Using GIS to compile and visualize daily management data in new ways enables innovative solutions for extraordinary situations.

8

DIGITAL PHOTOGRAMMETRY AND REMOTE SENSING

8.1 INTRODUCTION

Geographic information systems use many types of images, including aerial photographs and remotely sensed images. Images provide a rich source of raw information that can be digitized or scanned for use in a GIS and later analyzed. They also serve as backdrops in GISs. This provides a means to identify features for on-screen digitizing. Aerial photographs are usually taken from airborne platforms, although balloons and kites have been used. The capture, analysis, and interpretation of aerial photographs collectively is called *photogrammetry*. Remotely sensed images are also taken from airborne platforms, primarily satellites. More recently, other near-Earth platforms involving the use of airplanes have incorporated remote sensing techniques and methods. In addition, high-definition imaging (HDI), light ranging and detection (LIDAR), color infrared (CIR), and multispectral scanners (MSSs) together with radar are being used to generate images of Earth. Many new satellites have recently been launched that provide high resolution: *Ikonos, Quickbird, SPOT-5, Radarsat, Landsat, Orbview, EROS, IRS,* and *Cartosat,* among others.

There has been a shift from traditional analog aerial photographic emulsion and silver oxide–based imagery toward digital imagery, termed *digital photogrammetry* or *soft-copy photogrammetry.* New cameras and imaging techniques designed to process those images has been developed. In addition, software and hardware suitable for display and image handling have also been developed. Digital imagery can be found in both aerial photographs and remotely sensed images derived from satellite platforms. Imagery may be in either black and white or color. Color images allow for greater analysis than black and white, but their file size is greater also.

8.2 PHOTOGRAMMETRY

Aerial photographs are acquired from numerous perspectives. A *vertical perspective* is a photograph acquired from airborne platforms that have less than 3° tilt from vertical (Figure 8.1). These photographs do not vary by more than 1.5° from either side of a reference point perpendicular to Earth; they appear to have been taken straight down. The point extending from the center of the camera to the ground is called the *nadir.* The *principal point* is the geometric center of the aerial photograph, regardless of tilt. If there is no tilt, a planimetric view of the landscape is visible. This is analogous to the planimetric map perspective. A point halfway between the principal point and the nadir is referred to as the *isocenter.*

An aerial camera may also be tilted from a *perpendicular perspective.* Images that tilt up to 45° from the vertical perspective result in a *low oblique angle photograph* showing an area directly below as well as an area off in the distance; they do not usually show the sky (Figure 8.2). If the aerial camera is tilted beyond 45°, a *high oblique angle photograph* is captured and represented. The sky, or a portion of it, is generally visible in a high oblique angle photograph (Figure 8.3).

Most aerial photographs are 23 × 23 cm. in size, although the size may vary. These images can be seen stereoscopically, which permits a 3D view of the features observed. That is accomplished by overlapping the images both at the sides, along the

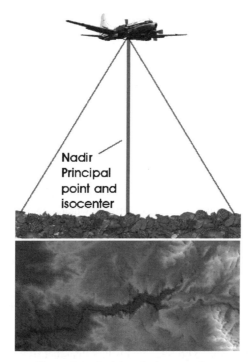

Figure 8.1 Aerial vertical perspective.

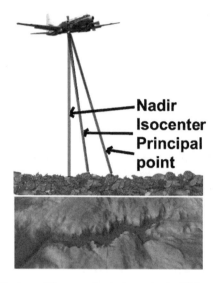

**Nadir
Isocenter
Principal
point**

Figure 8.2   Low-oblique-angle photography (<45° from center).

flight path, and end to end, or perpendicular to the flight path. Normally, images are side-lapped 60 percent and are end-lapped 30 percent. Again, the overlaps may vary for other reasons, which we will discuss shortly. Accordingly, higher amounts of overlap will require that more images be taken for a given area since each successive image along the flight path has less forward advance between exposures. This can be very expensive where large areas are involved.

Aircraft are flown along flight lines that run parallel to each other across the area to be photographed. The aerial photographs themselves are numbered consecutively for each exposure. This provides the means to order and manage the images themselves, which is very important for data retrieval and archiving and for developing metadata. Several thousand photographs may be taken for a given project, depending on the size of the area and the amount of overlap. GPS and inertial navigation systems have become more popular in recent years, providing the means to navigate aircraft along selected flight lines. This has helped to ensure consistently spaced flight lines, and GPS position can be used to determine when to take a picture or exposure.

The photo scale measures the ratio between any two points on the photo and the same two points on the ground. The photo scale is very important to know because without a scale landscape, objects and distances cannot be measured accurately. An airplane can fly either above *mean sea level* (MSL) or *above mean ground level* (AGL). Using AGL, a basis for determining photogrammetric scale can therefore affect that scale and subsequent measurements. Scale is usually derived on an AGL basis. Thus, the aircraft flies above MSL at a constant altitude, but the distance between the aircraft (and camera) and the ground varies—AGL. When flying over high mountainous areas, the distance above ground level is less than that flying over valleys.

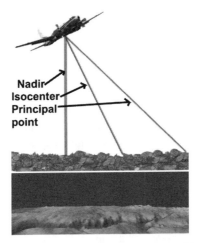

Figure 8.3 High-oblique-angle photography (>45° from center).

The principal point is the point on the ground from which the flying height is obtained; it is the center of the aerial photograph. The nadir and principal point coincide when a vertical nontilted photograph is taken. If the aircraft tilts, the nadir remains directly below, at a point perpendicular to the ground. However, the principal point of the aerial image changes, moving farther from the nadir with respect to the tilt angle.

Scale is determined from aerial photographs based on the flying height and focal length of the camera (see Table 8.1):

$$\text{scale} = \frac{\text{focal length}}{\text{flying height (above ground level)}}$$

Table 8.1 Photogrammetry scale by height and camera focal length

Aircraft Altitude (feet AGL)	Focal Length	
	f/150 mm	*f*/300 mm
500	3,333	1,667
1,000	6,667	3,333
1,500	10,000	5,000
2,000	13,333	6,667
2,500	16,667	8,333
3,000	20,000	10,000
4,000	26,667	13,333
5,000	33,333	16,667

Assuming that an aircraft is flying 2000 m above sea level and the focal length is a standard 152.4 mm (6 in.),

$$\text{scale} = \frac{0.1542m}{2000 \text{ m}} = 1:13,123$$

Alternatively, there may be times when one wishes to acquire aerial photographs at a fixed scale. In such a case the focal length remains the same but the flying height is unknown and must be calculated. Let's assume that we want to have aerial photos with a scale of $1:13,123$. Then

$$\text{altitude} = \frac{1}{13,123} = \frac{0.1524 \text{ m}}{x \text{ altitude}}$$

$$x \text{ altitude} = 2000 \text{ m}$$

It can be seen that focal length, flying height, and scale are all interrelated and that varying one will vary the others.

It should be remembered that flying height is calculated from the principal point relative to the aircraft's altitude. On flat land this would mean that most of the ground would have a similar scale relative to the principal point. However, this is often not the case, due to the fact that terrain varies in height. It is for this reason that we refer to the photographic scale as the *nominal scale*—it is relative to the principal point only (Figure 8.4). If the terrain varies while the aircraft maintains a level flying height, those objects whose elevation are above the principal point will have a different scale. Similarly, those objects below the principal point, such as in valleys and

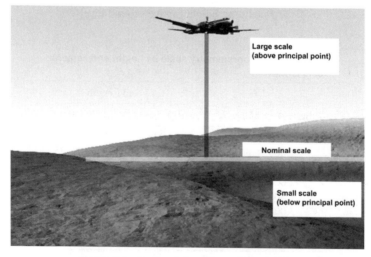

Figure 8.4   Scale relationships to flying height.

lower-lying areas, will have a different scale. In practice, then, most aerial photographs have very few locations with similar scale, due to topographic relief. Water bodies and other flat lands would have a similar scale across the photograph.

Another way to calculate scale is by the ratio of map scale to photo scale. Since a distance on the photo represents a distance on a map, then

$$\text{scale} = \frac{\text{photo scale}}{\text{map scale}} = \frac{\text{photo distance}}{\text{map distance}}$$

This can also be useful for calculating distances for a specific map scale. Assume that we have a series of aerial photos with a nominal scale of $1:40,000$. A city street is measured and found to have a distance of 30 cm. One might want to know the length of that street on a map with a scale of 1 mm = 150 m.:

$$\frac{1:40,000}{1:150,000} = \frac{30}{x} = 8 \text{ cm}$$

A primary advantage of aerial photography is that it may be viewed stereoscopically, providing a 3D representation. This is accomplished through the use of stereoscopy, a stereoscope being placed directly over two aerial photographs. Viewing two aerial photographs that were taken one after the other through the stereoscope results in a 3D representation of the entities on the stereo pair. Each aerial photograph will have a series of *fiducial marks* located at the corners of the photograph and at the four midpoints between corners (Figure 8.5). The orientation of aerial photographs is called *photo geometry.*

Extending a line from two opposite fiducial pairs on each photo will form the principal point (PP), the geometric center of the aerial photograph as viewed from the camera. The *conjugate principal point* (CPP) is the principal point from the adjacent aerial photograph. Each aerial photograph will have one principal point and two conjugate principle points, one from the adjacent photo to each side along the flight path. Through the stereoscope these conjugate principal points appear to be the principal points and to be floating on the adjacent image. They are located through the stereoscope and marked with a photo-marker or small pinmark.

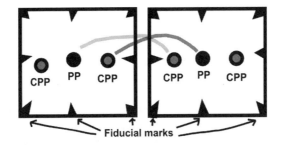

Figure 8.5 Photo geometry.

Drawing a line (CPP–PP–CPP–PP) through a series of aerial photographs identifies the flight line. It should be noted that the flight line does not necessarily always line up perpendicular to the axis of the image itself. The reason for this is that although the camera body itself is fixed in the aircraft, the aircraft may not be tracking a straight line, due to wind resulting in the aircraft rotating or changing attitude. The distance in the air from one exposure to the next exposure is called the *airbase.* Since that distance (more specifically, the camera center) corresponds with the principal points of two adjacent photos, we refer to the distance between PPs on the photos as the *photo base.* That distance can be measured on the stereo-pair using a fine ruler. Similarly, the airbase distance can be measured knowing the aircraft velocity and exposure rate between frames.

The ratio between the aircraft altitude and the airbase is known as the *base–height ratio* (BH). Larger BH ratios provide higher levels of map accuracy. The BH ratio can be calculated as follows:

$$\frac{B}{H} = \frac{\text{airbase}}{\text{altitude}} = \frac{\text{photo width} \times \dfrac{100 - \% \text{ overlap}}{100}}{\text{focal length}}$$

As can be seen, the focal length and flying height of an aircraft directly affect the base–height ratio. Sometimes there is a blurring of aerial photographic images. This can be caused by long shutter speeds while the aircraft is moving. One way to reduce blurring is to use faster shutter speeds; alternatively, decreasing aircraft forward velocity will decrease image blurring. More recently, *forward motion compensation* (FMC) has been developed. Pictures taken with FMC have less blur since the film itself is moved as the shutter is opened. A comparison of ground distances with image distances will provide an indication of *image motion.* Image motion can be affected by focal aircraft velocity, aircraft altitude, film speed, and shutter speed. High film speeds are more sensitive, requiring the camera shutter to be open for a shorter time.

Areas can be calculated from aerial photographs using a dot grid or line grid. A *dot grid* consists of a clear sheet of acetate on which a series of equally spaced dots are marked. A *line grid* consists of a clear sheet of acetate with a series of equally spaced parallel lines. Using a dot grid, assume that you have a photo scale of 1 : 20,000 and there are 20 dots/cm^2. Thus,

$$1 \text{ cm} = 20,000 \text{ cm} \quad \text{or} \quad 200 \text{ m}$$

$$1 \text{ cm}^2 = 400,000 \text{ m}^2 = 40 \text{ ha}^2$$

$$\frac{40}{\text{cm}^2} \times \frac{\text{cm}^2}{20 \text{ grid dot}} = 2 \text{ ha/dot}$$

Using a dot grid is simple. Toss the grid over the aerial photograph and count the dots. For those dots on the edge of the area or object being measured, count every second dot. Then multiply by the area per dot. Repeating this procedure a few times, then taking the average count, will provide a fairly accurate measurement of area. Keep in mind that the photo scale is relative to the principal point. Therefore, those objects closer to the plane (where the principal point lies in elevation) and extending through the image will be more accurate, the scale varying based on relief and the height of the objects themselves.

A line grid is used in a similar manner. Assume a scale of $1:25,000$ and a line grid measuring 20×20 cm, with the lines 1 cm apart (20 total). Then

$$1 \text{ cm} = 25,000 \text{ cm} \quad \text{or} \quad 250 \text{ m}$$

$$20 \text{ cm} \times 250 \text{ m} = 5000 \text{ m}$$

$$\frac{5000^2 \times \text{m}^2}{10,000 \text{ m}^2/\text{ha}} = 2500 \text{ ha}$$

$$20 \text{ grid lines} \times 20 \text{ cm}/\text{line} = 400 \text{ cm}$$

Assume that the object being measured with the line grid has a total run length of 19 cm. As an example, you are measuring a city block and within the edges of the city block you measure the grid lines on the acetate within the area, totaling the length. Then

$$\frac{2500 \text{ ha}}{400 \text{ cm}} \times \frac{x}{19 \text{ cm}} = 118.75 \text{ ha (size of the city block)}$$

These are very practical calculations for handling aerial handling photographs. GIS professionals can use them effectively when setting the scale for backdrop images, before on-screen digitizing. They are also useful for those interested in developing visualizations where 2D and 3D objects need to be scaled based on images.

There are photogrammetry standards established by the American Society of Photogrammetry for mapping and the creation of map products. Many mapping products originate from aerial photographs, including cadastral maps, topographic maps, ortho photos, and base maps. For creating contour maps from aerial imagery, the standards state that:

- Ninety percent of elevations determined from solid lines must fall within one-half a contour interval and the balance within one full contour interval.
- Ninety percent of spot elevations must fall within one-fourth a contour interval and the balance within one-half a contour interval.
- Ninety percent of planimetric features must fall within 20 in. of the contour interval and the balance within 40 in.

8.3 MOSAICS AND ORTHO PHOTOS

There are many types of mosaics and ortho photographs. A simple aerial photo mosaic consists of two, preferably more, aerial photographs that are arranged together to provide one larger view. These may include both side laps and end laps. There are three types of mosaics: ortho-photo mosaics, controlled mosaics, and uncontrolled mosaics. The most expensive, labor intensive, and useful of these three is the ortho-photo mosaic because they have been carefully cropped and rectified, and the relief and tilt displacement have been processed. This type of mosaic looks like a large photograph and can be used as a base map. All distances on this mosaic are accurate. All objects appear upright with no displacement away from the center of the image since it is truly vertical in perspective.

An ortho photo is a processed product, resulting in a raster digital image. This compares with a simple mosaic that does not exhibit the rasters, which can readily be seen when looking at the image close up. It should be noted that ortho-photographs are also color balanced, ensuring an even tone throughout the image, whereas simple photo mosaics are not. The controlled mosaics are those where the center portion (near the principal point) of the aerial photos have been clipped, then fitted together to form a mosaic. Since the amount of displacement is reduced in the center of the image they form more highly accurate mosaics, particularly if derived from terrain that is flat. Keep in mind that this technique can result in the end laps being lost, due to the cutting process, and can result in gaps. Therefore, flight lines with closer spacing and more end lap will prevent this from happening when constructing these types of mosaics.

Uncontrolled mosaics are simple mosaics that have not been processed for tilt or displacement. The aerial photographs are matched together to form one larger picture. These mosaics will appear uneven, often tone mismatched, and are useful largely for visualization of a larger area. Another type of uncontrolled mosaic is the *photo index sheet,* a series of aerial photographs matched along the flight lines and printed at a lower scale to form one larger image. These can look very peculiar because each photo of the mosaic is displaced from the principal point.

8.4   IMAGE REGISTRATION

In keeping with mapping and accuracy standards, as well as being able to use aerial images in a GIS, the images will need to be georeferenced. Mapping standards provide the criteria by which aerial photos are used to produce spatial products. Those products can be hardcopy maps or thematic layers for use in a GIS. As mentioned previously, contour intervals derived from photographs must meet defined criteria. These include minimum accuracy standards that relate to the methods used to acquire and process the photographs. These include target scales that provide an indication for the nominal scale for which the images are being developed and processed. Most spatial data are collected at a specific scale. Once the scale of the aerial photograph is changed using softcopy or digital processing, some of the information may be lost. This is particularly true where images are used in GIS. In these cases, rubber sheeting, the process of moving image boundaries to cover areas, also distorts the intended scale for which the original photo was produced. Mapping requirements also need to be considered. Such standards ensure that specific information in the aerial photograph must be visible and useful if it is to be used for specific purposes. In the last case it would not, for example, be useful to have aerial photographs at a scale of $(1:30,000)$ if one is interested in identifying distances between fire hydrants. On the other hand, at this scale it is reasonable to expect that useful forestry information could be obtained and adequately delineated, larger bodies of water evaluated, and highway networks represented.

A new aerial photograph has no coordinates or extents. It is simply an aerial photograph, probably in 23 × 23 cm format. It may be in black and white or in color. There is no way to determine distances on an aerial photograph unless the scale is known. The scale provides the means to calculate areal distances. However, simply knowing scale does not provide an indication about where a house or car or boat is in relation to the real world. For objects to be related to each other, they must have coordinates. The image must be able to be represented on a planar surface, similar to a map, and coordinates ensure that one image can be related to other images and spatial data. Those coordinates (x,y) can be provided through any number of coordinate projections. These include, for example, an equal-area map projection, discussed in Chapter 3. There are advantages and disadvantages to each type of projection. Nevertheless, an aerial photograph must be georeferenced to a projection; we need to know the latitude/longitude or UTM coordinate of each feature in the aerial photo.

Without real-world coordinates or georeferencing, the image may still be used in a GIS but probably will not be useful as a thematic layer with thematic data already registered. Perhaps you have had the experience of bringing a thematic layer into a screen, then a backdrop image on the same scene, only to find that the image does not lie behind the first thematic layer. That is because the aerial photograph in this case is not georeferenced similar to the first layer. Both layers are using different coordinates, one real world and the other (perhaps) standard computer display coordinates. Every mapping system identifies digital pixels based on a grid. There is a coordinate for each pixel, and a projection has a map coordinate system. Standard computer display coordinates begin at a point in the corner (usually, the lower left) that is (0,0) and extend through the photograph, assigning each pixel a coordinate. The coordinate in the upper left might be (0,100), the lower right is (100,0), and the upper right (100,100). The number of pixels assigned to each image depends on the resolution of the image.

To construct a digital ortho photo, one of two methods may be used. A digital ortho photo may be created by scanning a diapositive transparency from an aerial photograph. Alternatively, digital ortho photos can be scanned from already existing non-digital products using optical–mechanical processing. The more accurate method is the diapositive method. The number of pixels in the digital ortho photo scanned depends on the scanning resolution, dots per square inch (dpi). Each pixel represents varying areal distance on the ground. Therefore, higher scanning resolution results in more pixels per unit area, therefore represents objects more accurately. Scanning resolutions can vary and are selected depending on the intended use of the ortho photo and the quality of the original aerial image. These pixels must then be georeferenced, assigning a unique coordinate to each pixel representative of the real world. The image must also have a datum and projection.

To georeference an aerial image, GPS can be used. By selecting a series of locations on the landscape using GPS, then recording them, a coordinate is created for each object or location. Ideally, the greater number of GPS ground control points taken, the better. The points selected as ground control points should cover the area of the entire aerial photograph. Often, nine control points are selected: three in the top, three in the middle, and three from the bottom of the image, and they are located in the field and their GPS coordinates recorded. The GPS data are later downloaded, the digital ortho-photo database is opened, and the points selected are assigned to the known locations in the data table. Once the ground control points are assigned in a GIS, all other points (pixels) will be registered, and the extents or boundaries of the aerial image known.

This sounds simple enough, but there are sometimes other issues to consider. The coordinates assigned may not in fact line up with the digital image. A reason that this may happen is due to the original scanning resolution of the diapositive. Alternatively, one may not be able to locate a ground control in the field accurately due to the resolution and scale of the aerial photograph. Finally, although highly accurate, GPSs still have errors, which we discuss Chapter 5. Other anomalies may contribute to inaccurate georeferencing, including using one datum in the GPS but a different datum in the GIS when assigning the ground control points. Both should use the same datum and projection. When the ground control points do not match the land locations, points may be moved slightly. This is usually accomplished using a GIS and is termed *rubber sheeting,* a process whereby a planar image or thematic layer is stretched and bent as needed until a flat accurate layer is created.

From an integration perspective it is important to realize that if many types of objects are collected in one day using the same GPS, this is preferred to integrating various sources of data over time. When data are integrated over time from different sources of technology and using different methods, more work is required to georeference independent layers. Some may require more or less rubber sheeting, depending on scale, accuracy, and the ground-truth method used. Taking one thematic data layer obtained at 1 : 10,000 and integrating it with data gathered at a scale of 1 : 30,000 is difficult to do. The data are optimized for the scales at which they were collected. To integrate them means a change in one or the other data scales, probably the smaller scale data.

The resolution of the aerial photograph is related to the camera lens used, motion related to the photo, and the film type itself. Generally, the scanning resolution used should be about one-third the size of the smallest detectable object that needs to be identified. Once all aerial images are georeferenced, they can be used to determine locations as well as used in GIS as thematic layers or backdrops. Keep in mind that the aerial photo is now in digital format. All pixels have coordinates (x,y) assigned. If heights are determined, each pixel may have an (x,y,z) coordinate. The data table will then have a value for each coordinate field.

8.5   AERIAL PHOTO INTERPRETATION

Aerial photographs are valuable for constructing planimetric maps and creating ortho photographs. They are also useful for performing aerial photo interpretation, the art and science of determining objects on an aerial photograph and gauging their significance. Each aerial photograph has numerous objects. They may be in black and white, or they may be in color. The objective of interpretation is to identify the objects in the aerial photo accurately and assign them values. This is an integrative process that depends not only on being able to see the various objects on the aerial photo but also to deduce the relationships between objects. Roads do not lead into rivers usually, but instead, lead into other roads. Certain species of trees do not grow in certain types of sites (i.e., wet or dry). Buildings are not likely to overlap, and most manhole covers are not in sidewalks (although in some cities that can be the case), but in roads and along curbs.

Most aerial photo interpreters practice their interpretative techniques for many years, and have knowledge of the landscapes and objects they are interpreting. In this way they can readily discern differences between objects and classify them accordingly. The ability to classify objects accurately requires a strong capability to see stereoscopic images in stereo (3D). Objects are usually classified based on more than one factor. Needless to say, classifying all brighter objects as buildings is incorrect, but if the brighter objects look somewhat fluffy and are evenly dispersed around square objects, one might determine that they are deciduous trees. Going one step further, if these objects were located at certain elevations in a forest, one might, through association with other species, determine that they are trembling aspen, oak, or maple trees. This process of interpretation can be deductive or inductive. That, in turn, is related to the interpreter's knowledge of the subject matter being interpreted.

Deductive reasoning is a logical process and is based on the reasoning that several individual factors can lead to the conclusion or identification. It is selective by way of looking at several factors, sequential because one factor can lead to the next, and it is also analytical because one factor is evaluated against the others. An example of this is picking up an aerial photo and seeing many houses in a treed area and a large area with similar tone. Deductively, one then reasons that the house is on higher land and the area with similar tone is probably a lake or body of water. Inductive reasoning leads to generalization. Using the example above: "Lakes on aerial photos look like. . . ."

The integration of these processes is sometimes referred to as *elements of interpretation*. It more specifically involves knowledge of individual objects related to those in the aerial photo under investigation and knowledge of how one of those objects relates to other objects within the photograph. Therefore, an aerial photo interpreter who understands forestry is likely to have more success evaluating and classifying aerial images related to a forested area than is one who is more familiar with interpreting archaeology from images. The classification of aerial photographs involves many factors, including pattern, tone, texture, shape, and size. It may also involve shadows in some cases. For example, if the sun is low in the sky, objects will cast shadows. Buildings will look square in the shadows cast, compared to fir trees, which are pyramid shaped and pointed. It should be remembered, however, that aerial photographs taken too early or late in the day would result in more shadows, which would saturate the image and make interpretation more difficult. Most aerial photography used for interpretation purposes is flown around midday. Photos do have time stamps and can readily be seen.

The tone of an image refers to the color. In a black-and-white image there are varying shades of gray. Each shade renders a different tone. Objects with noncontinuous and irregular surfaces usually cast different tones. Objects with similar surfaces that are continuous tend to cast only one or two tones. The ability to distinguish tonal differences is an important clue for identifying many objects. Texture refers to the amount of tonal differences in the image. Objects with fewer tonal differences will have a smoother texture (e.g., a lake, homogeneous vegetation). Objects with many tonal differences will have greater or coarser texture (e.g., mixed-wood forest, variable land use). As scale changes to become smaller, there is less texture; objects are generalized and tend to merge together. As scale increases, objects become more readily identifiable, and larger texture differences can be seen. It is for this reason that aerial photographs are being used increasingly as backdrops in GIS. Using the image as a backdrop on screen digitizing can be performed if there are significant textural and tonal differences. Using an aerial photograph that is small scale as a backdrop is less useful since the textures again tend to merge.

Patterns of objects also provide good interpretative clues. The pattern refers to the orientation or spatial relationships of the objects within the photograph. A continuous line of square objects is a good clue that one is looking at a row of houses or apartment buildings. A continuous line across an aerial photograph may provide an indication that there is a road. If the line does not change elevation rapidly, it may be interpreted as a railway. Stands of trees that are homogeneous and particularly even-aged will have a similar pattern. Mixed-wood stands will provide varying irregular patterns, suggesting that more than one species of trees is present. Shape, which may be in 2D or 3D, is also an interpretive element. It is easier to identify objects when viewed stereoscopically. However, many objects in a photo are readily identified in 2D, particularly by trained and skilled interpreters. The size of landscape objects is an important and easily identifiable clue. When the scale of an image is known, the actual sizes of objects can be interpreted quickly. In such a case, large buildings are seen clearly a compared to smaller houses. Larger trees are seen more clearly than smaller trees and the size of water bodies is discerned rapidly.

One of the most difficult aspects of aerial photo interpretation is being able to associate objects that are being interpreted. As mentioned previously, some knowledge of the subject matter will provide the interpreter with a sound understanding of what types of object relationships can be expected. Specific species of vegetation are known to grow at certain elevations and are often not found in wet, lower-lying areas. Pine trees, for example, prefer well-drained sites. Spruce trees can tolerate moderate amounts of soil moisture, while tamarack trees are often found in moist, low-lying areas. We do not often see cellular telephone towers in valleys, nor do we find radio towers behind tall buildings. They are usually well exposed and at higher elevations.

The ability to use each interpretive element—size, pattern, shape, tone, texture, and association—is related to the quality of the aerial image. Higher-resolution images provide more elements and include wider variation within the elements. The interpretation of photographs is a process. The process leading to successful classification begins by identifying and delineating those objects that are easiest to identify. It is a mistake to begin by identifying the most difficult objects that can be recognized. Using the first approach, easy objects are identified one after the other until eventually most of the aerial photo is interpreted. Using those objects the interpreter can then begin associating identified objects using a process of deduction. Objects that cannot be identified will require the use of aerial images of a different scale and or visiting the field to identify the object.

8.6 PHOTO INTERPRETATION AND GIS

Aerial photo interpretation is related directly to GIS. Through the process of classifying objects on the image, a data table is constructed. That table will contain both spatial and aspatial data—object location and information about the object. A simple data table for forestry purposes may only include the (x,y) coordinate as the spatial data and the tree species noted in a field as the aspatial data. This is useful but does not tell us much about the forest. Instead, it would be more useful to know the species, age, height, number, and elevation of trees in the forest. Each of these variables will become a field in a GIS data table. Since each is a field, they can be queried in relation to each other. How many spruce trees are greater than 15 m tall? How much area do they cover? How close are the pine trees to the river, and are there deciduous trees mixed in with them? Greater classification results in a greater ability to ask questions about the objects and their locations. That is the value of a GIS. A very simple way to construct an effective GIS data table is to separate the interpretive elements into individual fields.

Using this method, then, any combination of interpretive elements can be queried. This method results in a robust data table that can be used for other purposes. In the case of forestry operations, silvicultural data can be integrated and evaluated over time. Silviculture is the art and practice of growing trees—more specifically, identifying and studying the phenology of tree growth. With this data table a forester could interpret the aerial images this year, then reinterpret the images in 10 years. A comparison of the tables would then provide clues about the forest's change over time.

Developing similar tables for determining land use change over time can be accomplished using aerial photographs. In this case the interpreter, using the photographic elements identifies building, roads, and associated urban change patterns. These data may later be coupled to demographic information obtained by enumeration or may serve as the basis for modeling travel times and determining where new shopping malls and other services should be located. In urban applications, aerial imagery may also be integrated with engineering information pertaining to utilities, water, and city services.

8.7 REMOTE SENSORS

Satellites are providing spatial information of increasingly higher resolutions. Their resolution meets or exceeds the resolution of aerial photographs in many cases, depending on the satellite. The price of satellite imagery is often cheaper than the cost of aerial photographs. One satellite image may cover 100 km^2 or more, whereas in some cases it would take hundreds of aerial photos to cover a similar area. There are several advantages in using satellite imagery:

- Lower cost per unit area (particularly for large areas)
- Recurring availability (satellites continually orbit or remain geostationary)
- High resolutions (often approaching 1 m or better)
- Availability of analysis software
- Ease of data transfer
- Use in GIS

A major difference between satellite scanning systems and aerial photo-based systems is that satellite imagery does not use a film, whereas most photography does. Therefore, remotely sensed images are digital from the moment they are collected. Two of the most common scanners in use today are the multispectral and the thermal IR scanner. Both of these are termed *cross-track scanners*. Most remote sensors measure thermal radiation at a wavelength between 3.5 and 20 μm, although anywhere from 8 to 13 or 14 μm is more common.

Cross-track scanners scan the landscape surface from side to side perpendicular to the satellite's or airplane's flight direction. As a satellite moves forward the cross-track scanner is moving back and forth continuously. As it moves, the reflectance from objects on the landscape is recorded, called *passive sensing or scanning*. The sensor onboard the aircraft or satellite does not emit any signal; instead, the detectors sense the landscape's surface, absorbing light of varying wavelength, analogous to the handheld camera. The earliest cross-tract scanners were developed in the 1960s for military use. This method of obtaining remotely sensed images remains in most common use today.

Similar to aerial photographs, remotely sensed images cover areas of different size. The size of the area covered depends on the height of the detection sensors in relation to the surface being imaged, as well as the *field of view,* the viewing angle of the detector. This can be compared to using a telephoto or wide-angle lens in a regular handheld camera. The wider the field of view, the wider the area covered will be. As the aircraft or satellite moves, it can see larger or smaller areas. The area that is seen for any point in time is termed the *instantaneous field of view* (IFOV). A smaller IFOV will result in increased resolution with more detail. However, because the sensor is moving or sweeping across the track, smaller areas will result in less reflected radiation being detected. Highly sensitive or multiband detectors can increase the chances of detecting wider reflected radiation since they are designed purposely to collect radiation within specific bands. Single-band detectors are referred to as *panchromatic scanners*. Passive sensors may have one or many bands or detectors. Those scanners with multiple detectors are then referred to as simply as *multispectral scanners*.

8.8 IMAGING SATELLITES

The first Landsat multispectral scanners (MSSs) were capable of achieving 80-m resolution. The Landsat thematic mapper (TM) developed later was able to detect objects down to 30-m resolution. These scanners are able to measure near-infrared, thermal infrared, and radio wavelengths. The Landsat program began in 1972 under the guidance of the U.S. Department of Interior and the National Aeronautics and Space Administration (NASA).

Several Landsat satellites have been launched since the beginning of the program. *Landsat 1 and 2* (launched January 22, 1975) and *Landsat 3* (launched March 5, 1978) are no longer providing remotely data and have been decommissioned. *Landsat 4* (launched July 16, 1982), *Landsat 6* (launched in October 1993; failed to obtain orbit), and now *5* (launched March 1, 1984), *7* are providing data. *Landsat 1, 2, and 3* gathered MSS data; *Landsat 4* and *5* collected MSS and TM data. Numerous countries have launched satellites (Table 8.2). More recently, private corporations have en-

Table 8.2 Imagery satellites

Satellite	Owner	Year Launched	Resolution	Type
Landsat 1 and *2*	U.S.	1975	—	MSS
Landsat 3	U.S.	1978	—	MSS
Landsat 4	U.S.	1982	—	MSS, TM
Landsat 5	U.S.	1984	30 m/120 m	MSS, TM
Landsat 6	U.S.	1993	Failed	Failed
Landsat 7	U.S.	1999	15 m/25 m	TM, ETM, MSS
JERS	Japan	1992	15m/25 m	SAR
SPOT-1	France	1985	10 m/20 m	Panchromatic, MSS
SPOT-2	France	1990	10 m/20 m	Panchromatic, MSS
SPOT-3	France	1993	10 m/20 m	Panchromatic, MSS
SPOT-4	France	1998	10 m/20 m	Panchromatic, MSS
SPOT-5	France	2002	2.5 m/5 m	Panchromatic, MSS
IRS-1C	India	1995	5 m/20 m	Panchromatic, MSS
IRS-1D	India	1997	5 m/20 m	Panchromatic, MSS
ER-2	European Space	1995	30 m	SAR ASAR
Sovinformsputnik	Russia	1997	1 m/1.6 m	Panchromatic
Ikonos	Space Imaging, U.S.	1999	1 m/4 m	Panchromatic, MSS
EROS A-1	ImageSat, U.S.	2000	1 m/1.8 m	Panchromatic
Quickbird	Earthwatch, U.S.	2000	1 m/4 m	Panchromatic, MSS
Orbview	Orbimage, U.S.	2001	1 m/4 m	Panchromatic, MSS
Radarsat-1	Canada	1996	8 m	SAR
Terra-AM-1	U.S.	1999	250 m, 500 m, 1 km	MODIS
CBERS 1	Brazil/China	1999	20 m	MSS

tered into the field of providing satellite images and products and are launching their own satellites. The evolution of remote sensors has been rapid and has resulted in higher and higher resolutions. Resolutions for many of the satellites launched within the last two years are below 5 m.

The *Ikonos* imagery is available in stereoscopic format, as is *SPOT-5,* which also has a supermode. In this mode, one cross-track scanner makes the first image, resulting in a 5-m pixel resolution. A second scanner, trailing behind, is slightly overlapped to the first (similar to aerial photography) and produces a second image. Since there is an overlap of the images, a smaller 2.5-m pixel is produced (Petrie, 2001). A new satellite from India (*Cartosat*) will be launched in 2002 and provide 2.5-m resolution. A satellite owned by the Japanese government, *ALOS,* will also be launched and is expected to provide 2.5-m panchromatic images. Several governments and companies have satellites planned for launch in 2002 through 2005.

8.9 SATELLITE PRODUCTS

Since each satellite may utilize a different type of sensor, individual satellites are optimized for particular types of products and applications. In the case of *Landsat* imagery, several bands are used in the multispectral scanner. Each of these is optimized for landscape evaluation and detection, responding to reflected landscape energy. The relationship of one band to the next is referred to as *interband ratioing.* These bands are calibrated for optimum reflectance and color differencing (Avery and Berlin, 1992). The coarser *Landsat* products with 30- and 90-m resolution are designed for smaller-scale work. Consequently, they are often used in the study of agricultural lands, water networks, vegetation studies, and other investigations covering larger areas. Those products with higher resolution (<5 m) are often used to generate orthoimages and for the construction of digital elevation models. Radar imagery (*Radarsat*) is particularly useful for building digital elevation models of the landscape. The advanced very high resolution radiometer (AVHRR) satellites use a broadband four- or five-channel scanner. They can sense in the visible, near-infrared, and thermal infrared areas of the electromagnetic spectrum. Thus, they are very useful for vegetation studies (NOAA). *Landsat* TM images have been used extensively to study forests (Slaymaker et al., 1996).

One of the most interesting satellites recently launched was the shuttle radar topographic mission (SRTM). On February 11, 2000, the SRTM payload onboard the space shuttle *Endeavour* was launched into space. With its radars sweeping most of the land surfaces of Earth, SRTM acquired enough data during its 10 days of operation to obtain the most complete near-global high-resolution database of Earth's topography. To acquire topographic (elevation) data, the SRTM payload was outfitted with two radar antennas. One antenna was located in the shuttle's payload bay, the other on the end of a 60-m (200-f) mast that extended from the payload bay once the shuttle was in space. Images from this mission are now beginning to be distributed. The higher resolution formats (<90m) will become available later through special permission from the U.S. government. There are several suppliers of satellite imagery.

8.10   IDENTIFICATION AND DELINEATION

Satellite images can be obtained in either panchromatic or color formats. Color formats provide greater opportunity for image analysis. Image analysis involves the delineation of individual differences and their grouping from each other. Similar to aerial photo interpretation mentioned earlier, image analysis is concerned with pixel relationships (Grave 1999). These relationships are referred to generically as *image analysis*. The process of determining the differences may be conducted in numerous ways. Usually, a remotely sensed image was analyzed using image analysis software. These software packages are capable of importing, georeferencing, and displaying the images. They are also capable of interpreting pixel differences. This segmentation of differences may be simple; for example, it may only include color. Or it may be more complex, algorithms being used that are capable of determining groups of similar and dissimilar pixels. Additionally, many of these software packages are capable of performing interpolation of pixels. Similar pixels may be grouped together and assigned a similar classification.

Like aerial photography, the classification of remotely sensed images can involve information about the landscape; that is, certain reflectance or absorption in the case of active sensors such as radar will result in similarities between similar materials. However, it is not as straightforward as it sounds. Grass of a similar type in one place may not result in a similar reflectance from grass in a distance location. In this case, ground truthing is needed to ascertain the characteristics of the grass in a more detailed evaluation.

When remotely sensed images are classified, they may use supervised or unsupervised classification. *Supervised classification* involves calibrating the pixel values in the image with real notes and observations taken from the field. Using supervised classification, higher accuracy is obtained when determining and segmenting the image. *Unsupervised classification* is the process of delineating pixel relationships and values based on the information in the remotely sensed image. In such a case, all grass would be assumed to have a similar pixel value, as would all coniferous trees. In highly variable terrain with many pixel variations, unsupervised classification can be quite accurate. However, on landscapes that have similar pixel values, it is more difficult to determine differences between values—this is where visiting the site is advantageous. Then a supervised classification can be made. When we begin to associate image pixel values with specific landscape entities, we are constructing and developing training pixels. These pixels later serve as the basis for further delineation and classification of the images.

Interpolation or grouping of pixels can be accomplished using numerous methods, as discussed in Chapter 4. This is sometimes called *cluster analysis*. The simplest clustering approach is by distance, or spatially. Other methods look at the formations of the clusters and consider variation between the clusters. In such cases those clusters with high variation remain, while those with lower variability between clusters are integrated. Often, cluster size is used as a primary factor when segmenting remotely sensed images. The reason for this is quite simple; every individual pixel could be a group. This is not wholly unreasonable to consider, especially when pixel cell size may represent 90 m or more on the ground. However, it is highly unlikely to result in a useful classification system. But the analyst must begin with an understanding of minimum pixel resolution of the image and the minimum area that must be represented in the following classification.

A *dendrogram* provides a means of discerning differences between classes. They are usually constructed horizontally, and the length of the lines provides an indication of variability between classes. A quick look at a dendrogram pertaining to pixel classes can provide much information about the relationship between clusters. If the line lengths are similar, there is more equally distributed classification over the pixel range. Alternatively, if some of the lines are much longer or shorter than others, that usually indicates those pixel values that are being under- or overclassified. In effect, then, not enough pixel values are forming a cluster or group, or too many are, meaning that the class should be broken into smaller value ranges. Ultimately, the classified satellite image should be developed with some understanding of the final product or application use. While an image could be segmented and 50 classifications developed, it does not necessarily follow that those classifications can be used to make an easy-to-read or easily understood map. Therefore, before beginning image classification, the image analyst must have some idea of what he or she is attempting to classify.

Classifying color itself has traditionally been the approach that most remote sensing analysts have used to segment images. There are other methods. One new method involves a study of the spatial relationships of individual pixels and their shape. This is something that photogrammetry has been doing for quite some time. Using this method, the spatial relatedness of pixel values is determined. This then leads to inclusion of shape. Buildings will have pixel values that appear to form lines. They will be near other pixel values that also form lines. Agricultural areas will not only have similar colors or values but also tend to be associated with other agricultural crops, which have their own values. Often, agricultural areas will have fences or tree rows bordering them. These will generate their own unique reflectance that satellite sensors will record. When associated with agricultural crops, these objects will serve to delineate and possibly classify the extents of specific agricultural areas.

EXERCISES

8.1. What is meant by the term *photo geometry?*

8.2. For what purposes can ortho photos be used? Provide two examples.

8.3. How is image registration accomplished? Outline the steps you would take to georeference a series of aerial photographs.

8.4. There are several available sources of satellite imagery. What are some of the considerations in selecting one product rather than another?

8.5. Briefly speak about the value of stereoscopy indicating how it can be used for building topographic maps suitable for GIS analysis.

8.6. Most aerial photographs can be interpreted. Name the interpretive elements that could be used in interpretation.

8.7. Are oblique-angle photographs useful for any applications? If so, name them and describe them.

8.8. Scale can be determined using two different methods. Discuss these methods and provide examples of your work.

8.9. A series of aerial photographs is called a *mosaic*. Can a mosaic be analyzed in a GIS? How would you go about this process?

8.10. What are the advantages and disadvantages of aerial photography compared to remotely sensed images from satellites?

8.11. Scanning is the process of converting a raster image to digital format. Explain the considerations when scanning aerial photographs for inclusion into GIS.

8.12. What is the importance of base–height ratio (*B/H*), and how does that relate to interpretation?

9

VISUALIZATION

9.1 INTRODUCTION

Defining visualization from a geotechnology perspective is a unique challenge. The challenge for geotechnology professionals in defining visualization is complex due to the fact that so many disciplines are involved in geosciences and spatial information services in an integrated fashion. Visualization has been referred to as visualization, geovisualization, scientific visualization, landscape visualization, cartographic visualization and a host of other terms. The *Oxford Dictionary* defines visualize as follows:

Visualize. v. to form a mental picture of.

The concept of visualization is no clearer from a geotechnology standpoint, and visualization may occur in many forms:

- Hardcopy map
- Graph, chart, text, or table
- Computer display in 2D, 3D, or 4D
- Aerial photos, satellite images, LIDAR, or image
- Theme, layer, or coverage

Those involved in geoscience will use many of these individually or together. A fundamental element of visualization using geotechnologies is the database. It is through the use and application of geotechnology that data tables (databases) are created. Data from these tables are used to build topology, which is then used in either a vector or a raster model form. Whereas visualization has evolved from computer graphics, GIS has evolved from the computer database (Thurston et al., 2001a, b). GPS data, which are vector data, can be downloaded and analyzed without actually using a GIS where data points or waypoints can be rendered in graph form using a spreadsheet—the plotting of points forming lines and polygons. Instrumentation can be used to collect meteorological, hydrological, light, sound, density, and other information and tables constructed on the basis of time, indicating change over time, all without using GIS. The analysis of data tables is something a GIS is uniquely capable of doing, in addition to rendering the results in visual form. Many 3D sets of data (x,y,z) are often rendered in 2D. In fact, one could argue that almost all spatial data are rendered in 2D except for those cases where true 3D monitors and immersive environments are being used. Does it follow, then, that 3D rendered in 2D is only two-thirds visualized?

Virtual reality is becoming a popular tool in visualizing 3D GIS data. Direct interaction with the GIS data, however, is often limited (Germs et al., 1999). Do restrictions in technology that do not allow for large 4D data sets to be rendered interactively (with the exception of supercomputing) mean that we cannot visualize them in our minds? Many of us have said to another person: "Paint a picture in your mind." Is that visualization? Is it possible to construct a map in text form, providing only words and numerals, so that another can visualize a map in their own minds? Imagine a map with the words *forest, road, house,* and *lake* on it. Is that any less meaningful in forming an image in one's mind? How much information would it take to do that? What are the considerations in doing that? Or is it a matter of providing text, tables, graphs, or a picture before visualization can begin? Part of the issue in defining visualization is that something does not have to be viewed to be visualized. Long ago the field of exploratory data analysis (EDA) was developed (Turkey, 1977), which emphasizes the seeking of unexpected structure and development of descriptions through graphic summary, robust statistics, and model fit indicators—in other words, exploration.

Maps, thematic layers, and other output rendered from geotechnology all provide visuals, but that is different from visualization. Visualization is an active process which engages the viewer who utilizes and interacts with visuals. Whereas three- and four-dimensional GISs have been developed to represent multidimensional geophenomena using geometry, the development of spatial multimedia and virtual reality systems has opened up new possibilities for multidimensional representation of a more direct nature (Camara and Raper, 1999). Although cartographers have been aware of developments taking place in scientific visualization, they have not been able to agree on how the term *visualization* should be used in the field of cartography (Slocum, 1999).

The term *visualization* in geography has increasingly become disjointed, selective, and confusing. The term *geographic visualization* by researchers (MacEachren, 1994; MacEachren and Kraak, 1997) describes visualizations that are geographic in nature. The term *geovisualization,* implying a connection to geography, covers a wide territory, since everything is located in time and space. Does it follow, then, that a video of the neighborhood is a geovisualization? *Cartographic visualization* would seem to imply only those applications involving cartography, but what about charts, tables, and text that are also connected directly to making the cartographic product? Are pictures on GPS liquid-crystal displays visualizations? They can and do show geographic content—rather accurately at that. Does a virtual reality presentation of landform and other geographic information qualify as a visualization? They are not often based on reality, although often used, particularly in the entertainment, game, and fictional realms.

A GIS may have a broad set of data input, processing, and output capabilities, but GISs often lack 3D visualization and certain modeling functions, whereas specialized object-oriented packages designed for visualization and modeling can lack many of the other capabilities of a GIS (Kuiper et al., 1996). Visuals can be created from historical, current, and real-time spatial information. There is no time delineation with respect to creating visuals, and some are created using real-time data gathered from instrumentation at many distant locations.

The usefulness of visuals may vary, however. Mission-critical applications requiring quickly updated spatial information require prompt visual expression. That expression can be in the form of a table, text, graph, chart, or picture. Each of these serves to provide the end user with the opportunity to form an image through interaction (visualization). Visualization can occur in color or in black and white. Once the image is formed in the viewer's mind, the question becomes: Is it accurate? Did the visuals present the information in such a manner that they communicated their purpose? Or were the visuals presented in such a manner that they left questions unanswered or even created more questions—GISs are notable for doing that. Thus the elements of communication and exploration are part of the visualization process, and that causes GIS professionals to generate more thematic layers and to perform additional analysis. Not all people perceive similarly; perception is a cognitive process that varies considerably within the population and is affected by numerous factors. Therefore, the visualization process can never be sure that the purpose of a visual is received accurately or used as intended. Visualizations for geotechnology professionals are oriented primarily toward reality.

The objective and challenge for geotechnology is to use many methods and techniques to construct visuals that reduce the level of erroneous perceptions, thereby increasing the likelihood of effective communication. To do this, data tables of real information are accessed, processed, and presented. They seek to create visuals that represent real situations through time and space. Inclusion of a legend on a map would be an example, as would scale. Yet other pieces of information cannot adequately engage the viewer in the visualization process effectively, due to constraints.

Errors in the data may arise from constraints in computing. This would include the technical inability to render images at the level of detail needed to portray the visual. Who is a *visualizor* and who is a *visualizee* when viewing visuals? It would be hard to argue that a person working on a GIS analyzing thematic content is not both at the same time. Similarly, it would be difficult to argue that a person visualizing spatial information is not, in fact, engaged in both simultaneously—viewing and forming new visuals in mind and recording them (e.g., using a GPS). Thus, the visualization process can occur both asynchronously and synchronously. Visualization for geotechnology professionals could be defined as follows: "Geotechnology professionals apply geotechnologies to capture, manage, and analyze spatial and aspatial information for the purpose of presenting visuals based in reality, using geotechnologies that engage viewers interactively for the purposes of communication and exploration, while also creating new images and information of the same types." Clearly, geotechnologies are directed toward the creation of visuals, and the visualization process does not end but can result in new information being created (Figure 9.1).

The communication and exploration of information may at times lead back to the collection of other data, beginning the visualization process over and over again or incorporating alternate geotechnologies. What is particularly interesting about visualization for the geotechnology professional is that it engages the creator of spatial information early in the process. Data collection procedures may be planned for and consider the desired map or image to be viewed. GPS mission planning software may be utilized to organize data structures based on analysis procedures and capabilities in GIS. Aerial photographs can be planned for which take flying height into consideration, thereby ensuring the appropriate scale for delineating and interpreting them, which directly affects not only presentation but also potential interpretation. The use of color has major impacts and affects the types of equipment used and how it is applied, as well as economic considerations. Generally, those technologies that produce black-and-white products are less complex to operate, cheaper, and often as useful as color imagery for interpretation.

The operation of continuous instrumentation may, however, be costly. GPS data collection is not usually associated with color—other than the interpretation of values for attributes in the field. GIS have historically been able to represent classes of

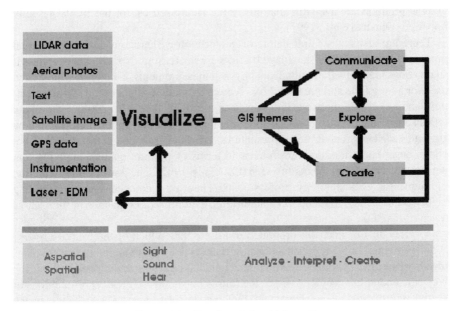

Figure 9.1 Creation of visual information.

objects through the use of color. In fact, many GIS now incorporate color ramps of predetermined shades for topographic, vegetative, and urban applications, taking into account aesthetics and attempting to assign map coloring more logically depending on the object or subject in view.

The assigning of shades by class is an attempt to render GIS content more adequately and realistically in 2D and 3D. Shading also has applications in 3D topologies with respect to a light source (e.g., sun, moon). Instrumentation also involves color for some applications. The measurement of aerosols in the atmosphere has utilized LIDAR and color gas chromatography to determine content levels. The visualization of information is necessarily affected by scale, and therefore generalization issues remain important considerations. Multiscale representation, an issue of growing interest and importance in geographic information systems, considers the spatial representation of entities at different resolutions in one common information system or thematic layer. Such representations have multiple benefits: for example, the transition from one scale to another, especially the use of coarse-to-fine approaches for data analysis (Sester, 1999). Objects and events apparent at one scale may or may not be as apparent during visualization at another scale.

The GIS professional and the cartographer use cartographic skills to render visualizations with higher or lower levels of detail as well as to create end products that are easier to understand, interpret, and utilize. For nominal data, objects should be distinguishably different, but since the data themselves are not ordered, there should be no perceptual ordering in the representation. For ordinal data, objects should be perceptually discriminable, but the ordering of the objects should be apparent in the representation (Rogowitz and Treinish, 1997). Maps are being developed that are directed to the uses of people with special needs (Birley and Tasker, 1997), and cartographic products are available that utilize the tactile senses, for use by the visually impaired (Ohtsuka et al., 1997).

There are also several translators of geotechnology hardcopy and digital visual products internationally—language barriers are decreasing. Not all cartographic visualizations are derived initially from physical measurements. *Cartograms* have been used for a long time and are useful for creating thematic content based on interpretations and perceptions between individuals or groups (Poiker et al., 2000). Schurmann (2001) and Gahegan (1998) speak about the ontology of perception, describing how attributes can be perceived and represented alternatively in topological structures. Yet others describe cartographic usefulness in terms of product quality, implying that perception is related to the quality seen (Couclelis, 1992). The issue of representation is unique for geotechnology professionals. They must contend with representing highly accurate 2D, 3D, and 4D spatial information in a 2D manner in most cases (even if it is 3D).

A series of four lines are presented in Figure 9.2. Which of the four are rivers? When presented in 3D representations for these types of data are more easily recognized because of landscape features and relief. Given that there is no legend, the viewer cannot quite decide which of the four is a river. Most VRML presentations of spatial information do not have a legend, nor do they have a scale. How can the viewer interpret this, then? It is almost impossible, but there is one clue—rivers cross

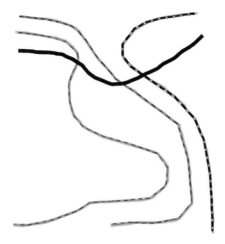

Figure 9.2   Perceptions of relief.

contours at right angles. For this image there may be one contour with three rivers crossing or three contours with one river crossing. Projections and datums create challenges for visualizing spatial information in the first instance and must be contended with during spatial data representation, taking into account geodesy before and during visualization. If maps of the province of Nova Scotia in Canada, the state of Florida in the United States, and a map of New Zealand or of Norway are presented, they all appear flat when you hold them or see them on a computer screen. Yet each is also different, due to datum and perhaps projection. They can be viewed independently, but try to place them together into a true 3D environment and comparisons between them are not readily possible, at least not without transformation to a common datum and projection—or reference. The simplest method, and indeed the most common means for representing spatial data, uses 2D representation.

9.2   TWO-DIMENSIONAL VISUALIZATION

Most maps are planar or planimetric projections. All objects appear in their correct positions horizontally as viewed from overhead. They can be used to measure distances accurately, are flat, and usually have a scale and legend. Aerial photos and satellite images have a central or perspective projection, where the center of the image is correct but objects are displaced radially as distance increases away from the center or nadir. Distances cannot be measured accurately on a perspective projection, due to displacement. Ortho maps do not have relief displacement, whereas photos do.

Planimetric representations of landscapes are used in agricultural operations, environmental studies, city planning, hydrological, survey operations, urban network applications, and oceanography, among others, and have a long history in cartography. It is the planimetric projection that most geoprofessionals view as a thematic theme in a GIS or hardcopy map. Each theme lays atop another in a planimetric fashion, aligned and georeferenced. When an image is used as a backdrop in a GIS, it remains a perspective image until ortho-rectified, at which time it is considered more truly planar, that is, when displacement and distortion have been reduced in the image and all objects appear in their correct place. Distances and angles can then be measured accurately on the image.

The easiest way to determine if one is looking at an orthophotograph is to look closely at the orientation of tall objects—they appear from the top down—not displaced and small pixels of equal size are apparent in the image. Determination of 2D coordinates and their representation is of interest to those involved in computer cartography, CAD, and GIS.

How objects appear has not been as important to GIS users as where the information could be accurately located and represented, with rapid database query and spatial operator capability, until more recently. The increased application of draping satellite images and aerial photographs over GIS models is an attempt to bring higher levels of photo-realism to database-focused products such as GISs. The representation of 2D coordinates can be accomplished by plotting coordinates on an x- and a y-axis. That is the easy part; or is it? Even though they can be represented in 2D, in reality they remain three-dimensional, occurring at different elevations as well as through time, but have been processed for representation purposes.

The 2D representation of 3D objects imposes that henceforth, all the objects will be represented at the same elevation; there is no depth perception. In 2D one can move left, right, or diagonally, never in or out with respect to the map or display surface. Immediately, this creates problems in labeling and identifying the z coordinates for objects being represented in a 2D model. One issue is to be able to label, represent, and assign attributes and values to those attributes from others on the 2D surface in a distinctive way. The figure of the river and topographic lines is an example of that. To improve on that figure, a legend could have been added. A scale would have been useful, too, thereby allowing for the determination of areal distance. Without those pieces of information, the map is not useful since it cannot be interpreted properly. It cannot be used as a theme in a GIS properly. If another theme is added, how do we know they both conform to the same scale and area? There is no way of knowing. It cannot be analyzed because features were not identified, and it cannot be measured because there is no scale with accompanying units of measure.

An aerial photograph was taken of a ranch in western Canada (Figure 9.3). GPS data were collected using static sampling and DGPS for a number of irregularly spaced points (diamond-shaped dots), each having (x,y,z) coordinates. The GPS waypoints can be seen. Elevation ranges from 0 to 20 m for the waypoints points displayed. Using a GIS, contour intervals were generated from those points creating 1-m contour intervals, and contour index lines can be seen in the image in Figure 9.4. Although these 3D data exist and could be used for 3D modeling, we will work in 2D with the data set. This poses a number of difficulties if the elevation information is to be represented, due to the fact that each dot may represent a differing height. One method of representing the information would be to create classes for each of the elevations. *Classes* are used to categorize individual attributes, assigning them into

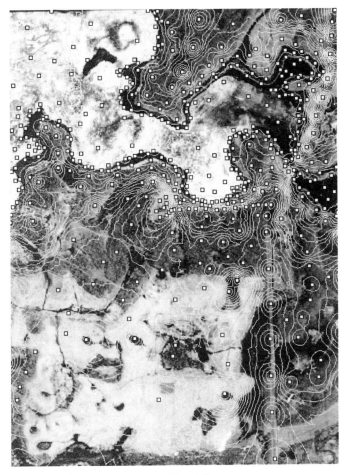

Figure 9.3   Ranch image.

groups. For example, if there are a group of 20 apartments in a city ranging in height from 100 to 200 m in elevation, they could be represented as one group (all elevations) or divided into five groups or classes based on their heights (i.e., 100–120, 121–140, 141–160, 161–180, and 181–200). They might also be placed into four classes based on the number of bedrooms in each apartment, or three classes grouped by number of inhabitants. Many combinations are possible, depending on how the data are to be represented and analyzed. Table 9.1 has 455 individual sample points where GPS data were gathered for the ranch. The table is divided into 20 classes, each representing 1 m. In other words, the elevations of the ranch area range from 0 to 20 m at the highest points. Twenty classes are quite a few, though. It is confusing to determine the differences between each class, and 1 m of elevation change is not all that easy to see sometimes. To make things easier, this was changed to five classes, for the GPS collected points and elevations were classed using a 1-m class interval (Figure 9.5).

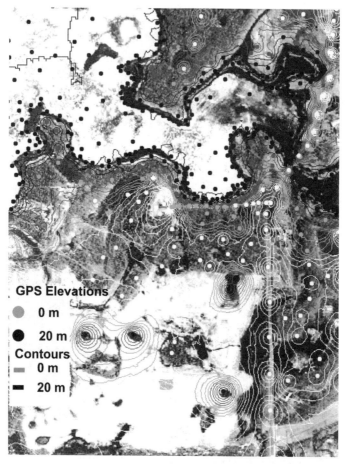

Figure 9.4 Landscape elevation classes.

The same method was used to classify the contour intervals. GPS sample points and contours are both represented. In practice, though, only one or the other need be represented to reduce redundancy. There are benefits in showing the contour intervals because they provide easily recognizable trends of elevation across the surface. Another way to represent the contours would be to label them, assigning the elevations. Computers represent images using a Cartesian coordinate system and a matrix of equally spaced pixels. The origin (0,0) of the coordinate system can vary from one software package to another. Some set image origin to begin in the lower left, while others set image origin in the upper left. Image size may vary depending on their dimensions and pixel resolution. Resolution is usually expressed in dots per square inch. A fine resolution image would have more dots per square inch of surface—1200 by 1200, for example. A course-resolution image would have fewer dots per square inch—480 × 320 dots or pixels.

For data at the same scale, one pixel represents a different distance for a fine-

Table 9.1   Landscape data

Elevation	Count	Elevation	Count
0.0	147	11.0	3
1.0	91	12.0	28
2.0	47	13.0	8
3.0	14	14.0	22
4.0	13	15.0	6
5.0	23	16.0	3
6.0	8	17.0	5
7.0	5	18.0	2
8.0	20	19.0	6
9.0	2	20.0	8
10.0	2		

Figure 9.5   Classed contours.

resolution image (higher) than for a coarse-resolution image (lower). Finer-resolution images are noticeable by the lack of sawtoothing or jaggedness apparent in the image. They appear clearer, smoother, and are easier to work with in ground truthing and digitizing functions. Coarse-resolution images appear to have small squares or pixels that are readily apparent to the eye and easily perceived. For 2D representation both coarse- and fine-resolution images can be represented. However, the distance a pixel represents is important to know for numerous reasons, including the ability to perform GIS spatial operations.

Data resolution and computer resolution are not the same thing. All computer-generated images, maps, and visual products are by definition raster based, utilizing a Cartesian coordinate system. They are projected onto computer screens or other surfaces using pixel coordinates. In Chapter 8 we discuss pixels and resolution further. Spatial information can occur in either a raster or a vector model. The capture of GPS data points for the ranch represents vector information, while the digital image backdrop for the figure represents raster information.

Data from each model can be merged. This can become tricky where widely different scales are being used. The aerial photography of the ranch had a scale of about (1:5000), meaning that 1 mm on the photo represents 5000 of the same units or 5 m on the ground. The GPS information was collected with an accuracy of 1 m. Therefore, no matter how much we want to try to maintain the 1-m resolution, it cannot be maintained. This is important to know since it can mean the difference between representing a house on land or in a lake. From a GIS perspective the GPS points can all be analyzed, the data table's queried, and new thematic layers created. A representation of the elevation change for the land surface can be represented.

The contours, image, and GPS data points can all be placed into a 3D viewer such as VRML and spun around, tilted, and viewed from various angles. But they remain in 2D, and some would even argue that 2D data should not be placed into 3D rendering environments but should instead viewed from a planar viewpoint—top down. Imagine that there is a set of data loggers collecting air temperature across a landscape simultaneously. At landscape and local scales, temperature patterns are strongly influenced by fine-scale insolation variation and micro-site factors, including vegetation, soil properties, and wind (Fu and Rich, 1999). The data logger sample interval is hourly, and there are 14 of them running in a 5000-hectare area. These monitoring stations are measuring vertical fluxes, and therefore each has an array of five sensors placed 2 m apart vertically (Figure 9.6).

After running for a month or two, the data are downloaded and compiled into a data table. Table 9.2 has been clipped to two sites to save space. The table itself represents all the information about air temperatures, causing the viewer to begin visualizing what the results look like in a picture. They could be graphed in a spreadsheet and shown as a bar, line, or pie chart, which would be easy to do. In the spreadsheet a new field could be created and entered for each of the (x,y) coordinates. In this way the 14 locations would be located and such questions developed as: At which stations within 1000 hectares and 6 m above the ground did the air temperature rise above 4°C between 0400 and 1100 hours? How do we begin to express that in 2D? Temperature, location, time, and height would all need to be considered in constructing the 2D vi-

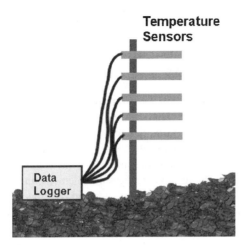

Figure 9.6   Temperature sensor arrays.

sualization. The answer to the question could be expressed as one map, but how are the other relationships represented for other locations, times, and temperatures relative to the first? One way is to class the variables as discussed earlier, using classes to reduce the numbers of individual maps. Another way might be to create an animation of the data for the 14 sites, either individually or one beside another. This can be achieved in a number of different ways.

Time change can be represented by a change in the background color, midnight being darkest and noon brightest; as time changes, the tone changes. Or it could be represented as numerals in one of the upper corners. As the time changes, the numbers change, where other variables comply and are represented with the time change. If there were five horizontal bars across the screen, each representing varying temperatures, they could change from blue (cold) to red (hot) as the time changed. The problem with this, though, is that it would be hard to see $1°$ changes and the screen

Table 9.2   Spatial temperature log, February 1, 2002

Time	Location 1					Location 2				
	1	2	3	4	5	1	2	3	4	5
0	2	3	2	4	3	2	3	3	2	5
100	3	4	4	3	4	3	4	4	4	3
200	4	4	6	5	4	4	4	4	5	4
300	5	5	6	5	4	5	3	6	5	6
400	6	7	6	7	6	6	6	6	7	7
500	7	6	7	8	7	7	7	7	8	8
...	...	...	...	...	...	...	...	...	...	...
...	...	...	...	...	...	...	...	...	...	...
2400	10	9	9	10	9	10	10	9	8	8

would appear as a kaleidoscope of changing color, while the background also changes color or numerals for the time change in the upper corners. Imagine 14 bar charts, each with five changing colors on your computer screen—definitely not a pleasing sight. This would be hard to interpret and very confusing. Did we mention that there is the choice of black and white or color for this—the choice is yours! By now it is readily apparent that the representation of spatial information can become very complex.

The commonest method for displaying spatial data is with a map—in the case of area data, choropleth and point symbol maps are the most common two forms. Choropleth maps average information for selected regions. Choropleth maps are commonly used to relationships for patterns in the data, where the word pattern normally implies regularities in the data. We discuss cartographic elements in our cartographic chapter. Patterns provide a means to separate out broad (or smooth) features of the data from local irregularities (or rough features). The problem is that the same data may be mapped many different ways, each yielding a different visual pattern (van der Schee and Jense, 1995).

Visualization is unique in that it allows for ways of representing information, although at times there are limitations to the effectiveness of the presentations. These can occur due to technical issues such as equipment and processing capability, or arise due to the method of representation—2D, 3D, or 4D. Add these to the issues of cartographic representation and things become more complex.

9.3 TWO DIMENSIONS AND INTEGRATION

When discussing 2D we are referring to either a visualization that is presented in a planar manner or the application of data that minimally involves a (x,y) coordinate, there is no such thing as 1D. GPSs are ideally suited for capturing 2D coordinates. This can be accomplished readily with any GPS navigator in point, line, or polygon mode. In fact, it is easier and quicker than the capture of 3D coordinates because it only requires the tracking of three satellites to record information. The GPS data tables for waypoint information in their simplest form will have three fields:

- Waypoint identification label
- x-coordinate
- y-coordinate

A 2D topology can be constructed for use in a GIS. Each of these coordinates can be represented and viewed in 2D. Two-dimensional information cannot be viewed in 3D. Although it can be presented in a 3D environment, it will look flat. The data may be exported to a GIS, where spatial analysis between points and times can be queried. There are not many complications in using 2D GPS data and representing them in 2D. Additional points, routes, and polygons can be appended to 2D GPS data tables, but the smallest identifiable unit will remain the point in a vector model. The simplicity of the data set is due to the fact all the information is spatial—the waypoint itself is a feature. In a 3D data set that is also based solely on spatial coordinates, it is also easier to represent. It becomes more complex when aspatial information is to be represented in either 2D or 3D. In GIS each 2D GPS point may be queried individually or together (i.e., show a feature with respect to another feature). There are no attributes present for the waypoints, nor are values present. Readers familiar with GPS know, however, that 3D data can be collected with a GPS, even those that only record waypoints in 2D. Consider the scenario where one knows that they will only be monitoring or measuring the same features in their GPS job for a period of time. The waypoints can be recorded as attributes instead of just calling them point 1, point 2, and so on—the feature is assumed.

Using manhole covers as an example, why not record their positions as either 0 or 1, where 0 means that it is not rusted and 1 indicates that it is rusted? Or, have four classes representing degree of infestation of grasshoppers, for example, where 0 is low, rising to 4, indicating 100 percent infested. When downloaded, the data table for these examples can later be appended in spreadsheet or GIS form to add the feature. Reviewing the earlier exercise for measuring the air temperature, assume that one rather than five air temperatures are being measured for each station.

For 2D each station could be located by coordinate and a color applied, depending on the air temperature class chosen. Alternatively, the numerical value for each temperature could be assigned to the station for any given time. Thus, a series of 24 maps would be rendered for a day, showing the hourly temperature for each. Analysis between stations could also be performed and thematic layers created. To reduce the numbers of maps, temperature data can be averaged over periods longer than 1 hour. Adding aspatial information to 2D coordinates will require consideration of two factors for 2D display:

- How to capture the entity with respect to its characteristics and processes (sampling interval, duration, and magnitude)
- How to represent the information captured simply without generalizing the information such that the characteristics are lost

A new version of VRML entitled Geo-VRML will include added cartographic capability, including the ability to scale, display coordinates, and generalize representation as the scale is changed. It is important to remember that 2D data will have only two spatial coordinates but may have many attributes and many values. Two-dimensional information can be derived from aerial photographs. Importing an aerial photograph into GIS will result in 2D coordinates. As the computer mouse is moved across the image, the changing (x,y) coordinates are seen. If LIDAR images or satellite images are imported into a GIS, they too will have coordinates based on the numbers of pixels in each image.

- 480×320 dots per inch $= 153,600$ coordinates
- 800×600 dots per inch $= 480,000$ coordinates
- 1024×1024 dots per inch $= 1,048,576$ coordinates

- 1200 × 1200 dots per inch = 1,440,000 coordinates
- 1600 × 1600 dots per inch = 2,560,000 coordinates

Naturally, it takes higher end graphics cards to render higher resolutions. There is a difference between zooming into an image at one resolution compared to increasing resolution of the image. If an image is zoomed in several times, it becomes blurry and almost difficult to see. An increase in resolution means that pixels are subdivided into smaller units; thus it takes considerably more zooming in on a higher-resolution image before clarity is lost. High resolution is clearer; low resolution is less clear. The drawback to resolution is the size of the data file. Images of higher resolution take considerably more space. In the case of Internet mapping, file size has major ramifications with respect to transfer rates. The point in this is that visualized images are based on the original image resolution compared to the presentation resolution.

9.4 GRAPHICS FORMATS

There are two formats that are used for computer graphics: bitmaps and vector graphics. Bitmapped images consist of a series of pixels, as mentioned previously. Usually, bitmaps are large and require a significant amount of space. Images using GIF (Graphics Interchange Format) or JPG (Joint Photographic Group) formats are compressed and save space. JPG images use a palette of millions of colors, making them suitable where subtle color changes are needed. There are numerous extensions for computer graphics:

- PNG (portable network graphics)
- .TIF [tagged image file format (.TIFF)]
- BMP [bitmap (Windows or OS/2)]
- .GIF [graphics interchange format]
- .JPG (.JPEG)
- .PCD (Kodak Photo CD)
- .PCX (Paintbrush)
- .TGA (Truevision Targa)

GIS images using a palette of 256 colors are ideally suited for graphics where subtle changes are not necessary, yet good color separation is required. The GIF formatted image also has the ability to render transparent colors, which is useful for showing underlying content. Subsequently, they also transfer over the Internet more quickly. Vector-mapped images are scalable and consist of lines, points, and polygons. They can be expanded or collapsed without losing clarity and can be printed on a print with higher definitions if the printer permits. Bitmap images cannot be printed at higher resolutions than exist in the original image. For visualizations where viewpoint and scaling change rapidly, vector graphics are widely preferred.

Often, users will take a large bitmap and scale it using the handles of the bounding box. This will cause the picture to look smaller, but the image size remains the same. In these cases it is better to resample the image to a smaller size, then render the image. Try taking a simple image and saving it in PNG, TIF, BMP, GIF, and JPG formats; then review each. You will see differences in each and also begin to see the loading time it takes between each type of image. The temptation is to take one format, especially JPG images, and alter them and save them into new files. Each alteration reduces the quality of the image, and this should not be done. Rather, the original JPG image should be placed in one of the other formats, altered, then saved into JPG format only prior to rendering. You have probably noticed that most GIS export to JPG or BMP formats. This means that all the themes created in GIS should be worked upon further as bitmaps and sent to a map server application only after they are turned back to JPG or other desired format. Subtle changes in colors, thus positions, and loss of apparent visible accuracy can result from continually flipping images between formats. TIFF format is not really one format but numerous formats appearing under the TIFF label. One of the issues in using TIFF file formats is that not all formats within a group can be seen using some software. This is why users will often hear others indicating that they have a TIFF but cannot see it.

The printing or plotting of images is another issue. Printers by necessity must rasterize or bitmap images for them to be printable. The plotter or printer cannot print a higher resolution than that which the image has, although it can print lower. Generally speaking, anything over 300 dots per inch is fairly clear, again depending on the subtle tone variations in the image. For GIS, most color ramps do not exceed 256 colors; therefore, any combination that uses a 256-color palette is sufficient. For taking higher-end visualizations and near-photo-realistic renderings, one needs to consider whether or not a GIS can actually produce the necessary resolutions.

This goes back to the issue of GIS being database focused whereas visualization is image or camera focused. It may be that visualization software should be used for those renderings requiring higher resolution, effects, and texturing—the GIS used primarily for generating spatial information with respect to the thematic layer. It should be kept in mind that a GIS that cannot convert between formats poses a problem, since the user may not be able to import or export with the GIS. Most major manufacturers of GISs have very good import/export functionality allowing for the following formats, sometimes many more:

- Arc Digitized Raster Graphics (ADRG; Military Specification-MIL-A-890007)
- BIL, BIP, and BSQ
- BMP
- Grid
- Erdas IMAGINE
- JPG (JPEG compression)
- RLC (run-length compressed)
- SunRaster files

- Tag Image File Format (TIFF) and LZW
- GeoTIFF
- DTED (Digital Terrain Elevation Data)
- IMG ERDAS
- GRASS (Geographical Resource Analysis Support System)
- DRG, DXF AutoCAD
- Topologically Integrated Geographic Encoding and Referencing System (TIGER)
- Vector Product Format (VPF)
- Digital Line Graph (DLG)
- Spatial Data Transfer Standard (SDTS)
- Shapefiles

9.5 THREE-DIMENSIONAL VISUALIZATION

Three-dimensional data include three coordinates (x,y,z). The third plane is usually the elevation for most data sets, but it does not have to be. It can, instead, be used to represent other information, such as temperature. Therefore, x-axis information can be either spatial or aspatial, when represented. When it is aspatial, magnitudes of measurements for a series of (x,y) coordinates are expressed. An example of this is charting a spreadsheet using a 3D histogram. However, 3D topology is based on the database and implies that (x,y,z) are points, lines, or polygons within the database to which other data refer. In most cases, DEMs are used for 3D purposes and may be combined with other data or thematic content. Aspatial information used in 3D, on the other hand, is generally used once, then discarded.

Therefore, a distinction is made between spatial data that are used to construct the database topology (i.e., physical features) and aspatial data that are simply represented in 3D (i.e., attributes, values), and 3D as we refer to it is a data model. Temporal GIS are systems for representing the temporal behavior of geophenomena when they have been projected from four dimensions to two spatial plus one temporal (Raper, 2000). Time becomes incorporated into most 3D GIS applications, and they are often referred to as being *spatiotemporal,* having both space and time. Although a series of 2D representations over time, indicating change would also be considered spatiotemporal. In database form, features, attributes, and values can and may change over time. What is most striking to the viewer about 3D as compared to 2D is that a higher level of realism is achieved. Viewing a 3D graphic provides the viewer with a sense of relatedness to the objects being viewed, incorporating perspective and depth, angle, and distance. These increase the level of interaction (or visualizing) with the representation. Communication can be increased using 3D, speeding up the image-forming process about the content being rendered. Whether or not that communication is effective depends on several factors related to what is being represented:

- 2D plus time, 2D minus time, 3D without time, 3D plus time
- Color apparent (hue, tone, and saturation)

- Period for which time is being represented (speed)
- Use of scale and legend
- Viewers' perspective in relation to objects being represented (angle and distance)
- Clarity of representation, computing hardware (including glasses or helmets)
- Level of knowledge of the user about the content

Ideally, the viewer would interact with the rendering through being able to change some of these parameters based on personal preference. The degree to which a visualization allows a viewer to interact for this purposes will probably lead to increased interest, communication, and further interaction. It is interesting to note the similarities between visualization for geography compared to video and computing games in this regard.

Higher realism and higher interaction equals increased involvement. Perhaps that is one of the goals for geographic visualization—to keep the viewer involved while the message is delivered. Digital elevation models are the most common form of visualization that geotechnology professionals view and utilize, and these can be represented as continuous surfaces with faces (Figure 9.7). Or, sometimes the lines between nodes are used for visualization purposes, permitting transparency to enhance visual effect (Figure 9.8). Each of these is a 3D surface, and aerial photographs and satellite images could be draped over them. Because these surfaces form topology, they can be queried further for:

- Slope angles
- Hill shading
- Aspect
- View sheds

Added objects and features can be integrated with the topology, and thematic layers can be assessed with respect to the physical surface, provided that the surface and thematic layers have similar georeferencing. Where additional objects are located

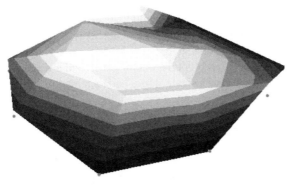

Figure 9.7 Digital elevation visualization.

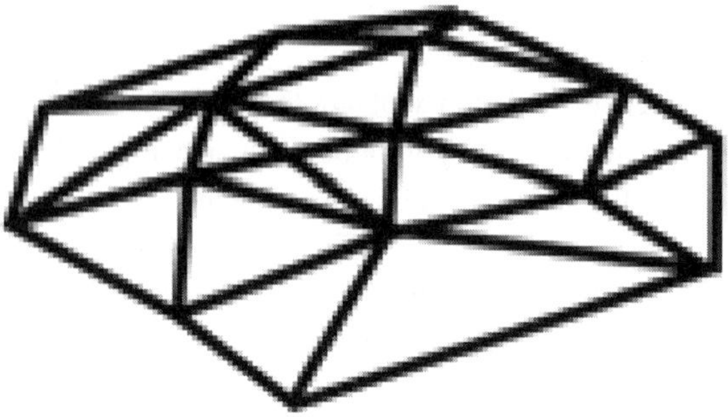

Figure 9.8 Visualizing transparency.

with respect to the surface, they can either be below, on, or above the z-coordinate (surface), but the (x,y) coordinates must conform to the coverage georeferencing. In this way, geological entities or subsoil attributes can be added and atmospheric events above the surface studied. The real value of 3D is that the viewer can see these phenomena with respect to the surface in their occurring position. There are other interesting perceptions that can be made with 3D representation. Using a conventional 2D approach, for example, human locations can be assessed from a different approach. In 2D the distance from one point to another may look very short; however, what is not seen is that there is a change in elevation of almost 200 m straight down. In 3D these points and the terrain between them are clearly evident. Consequently, the analysis of databases provides the answer for the shortest distance between two points and plots it fantastically in 2D, but in 3D it is doubtful that humans scale cliffs of this height with flat faces—unless they are mountain climbing.

The query could have taken the form of: "Show me the shortest distance between the two places where the elevation change is less than 10 m and the angle less than 40°," or something like that. But how would you know to ask the question before you began? Often, the person analyzing data in GIS is not the person who collected the information, so how are they expected to know what the field really looked like? Further, and perhaps equally interesting, how is the person to know the conditions over which humans can move? Do they make up their own parameters based on the height that they can climb? Very quickly, 3D would in this case identify that a large cliff exists. Having knowledge of that provides the opportunity to explore further and associate the terrain to other entities.

In fact, to ask the question properly, one might consider developing a model that includes the human capability to scale slopes and then integrate that information into the GIS as a model. Once done, any query relating to shortest paths for humans between two points would automatically invoke the model, which in turn would only provide the shortest distances between the two locations that humans are able to travel. The integration of models is not always a 3D process since it can involve 2D information.

In Chapters 3 and 5 we discuss surfaces with respect to GPS in detail. One of the interesting things about surfaces using GPS and GIS is that almost everyone looks at them from the top down. The data may be collected from air or ground platforms but usually ends up being respected from the top down. That has considerable merit, of course, where networks such as roads, rivers, houses, and streets are concerned, but for other types of information and phenomena, alternative methods are useful to assess phenomena from alternate planes, including the vertical plane.

9.6   VERTICAL GIS AND VISUALIZATION

Coupled to GIS and using vertical content, the rich rendering capabilities of visualization can enable analysis of geographic information while providing a broader range of rendering functionality. Analysis of visualizations can provide clues and greater understanding of landscape biodiversity indicators. Historically, ecologists have studied homogeneous regions to characterize and understand ecosystem processes and have avoided the heterogeneous areas between ecosystems (Fortin et al., 2000). Landscape management will require increasingly finer levels of resolution. The prerequisites for on-the-ground forest landscape management include a quantitative description of the forest landscape, a computer model, geographically prescribed harvest interventions, an understanding of spatial forest dynamics, and a GIS-based management design process (Baskent, 1999).

One study incorporated several geotechnologies for the purposes of assessing landscape biodiversity indicators quickly from the vertical plane (Thurston, 2001). The concept of vertical GIS is similar to conventional GIS applications applied from above. However, whereas top-down GIS covers very large areas horizontally, vertical GIS and visualization are focused on smaller localized areas. Vertical GIS is used to identify, assess, and monitor biodiversity indicators vertically and can be applied through time in either 2D or 3D. This is achieved using numerous geotechnologies, including GPS, image analysis, digital cameras, laser rangefinders, and GIS all coupled together. Visualizations are constructed vertically (side view) for landscapes and city areas (Figure 9.9).

The images are then analyzed using conventional image analysis techniques and classed using the iterative self-organizing data analysis technique (ISODATA). From those delineated classes, using TIN, a series of height fields based on vertices can be constructed. The images may also be analyzed using NDVI. Since the vertices are generated, they represent structure—they are indicators. A digital vegetation model (DVM) is then constructed that can be measured or analyzed by class: using overlay analysis and determining harmonic means or performing a spider analysis, for example. The classified differences for the side view of a section of forested land using the ISODATA are illustrated in Figure 9.10. The technique incorporates the following steps:

1. Determine if the project is large or small scale. Often, local investigation will permit direct application of vertical GIS.
2. Consider whether vertical information is going to be more useful than ae-

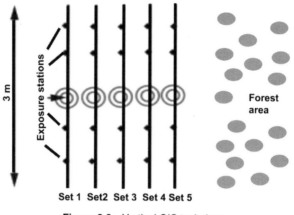

Figure 9.9   Vertical GIS technique.

rial photography, remotely sensed images, or other horizontally collected imagery. Is temporal monitoring going to be involved, requiring revisitation of coordinates?

3. If the project is small scale, consider initial use of aerial photography and remotely sensed images to identify sites that would benefit for more detailed investigation. Acquire geographic coordinates for later GPS input and navigation—vertical GIS control points.

4. Address the following elements: vertical GIS equipment selection; selection of digital cameras based on desired resolution; GPS equipment, preferably tied to a camera with coordinates; image analysis and GIS software; laser, transportation, personnel; computing speed and storage needs.

5. Input vertical GPS control points from predetermined horizontal imagery into GPS. Navigate to sites (consider climate and time of day and/or year).

6. Perform data collection, image capture, and laser measurements. Consider field of view and possible panoramic data capture from ground control points. Estimate GPS errors—consider static sampling.

7. Perform data download, with storage and backup.

8. Determine scales, consider ortho-rectification, compose photo mosaics, select unsupervised classification for physical measurements, construct stereo pairs, perform stereoscopic measurements, and determine training pixels for supervised classification if image analysis is involved.

9. Perform GIS analysis. Construct 2D and/or 3D visualizations.

10. Render by way of VRML, X3D, 3D Studio Max, CAVE, IRIS, VTK, or similar software.

In application, conventional top-down techniques for determining landscape biodiversity are used, and vertical visualization and GIS are applied if further investigation or data are needed at a more local scale. When GIS thematic themes are geo-

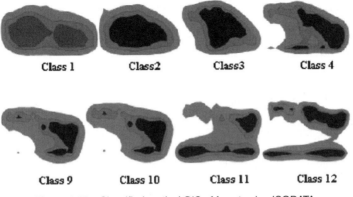

Class 1   Class2   Class3   Class 4

Class 9   Class 10   Class 11   Class 12

Figure 9.10   Classified vertical GIS of forest using ISODATA.

referenced, any point on the ground can be selected and its coordinate uploaded to a GPS navigator. The coordinate is then located, after which images can be located. The databases constructed from vertical images can later be integrated with other horizontal information. One set of data is large scale and represents one view perspective; the other set is small scale, representing another viewpoint—one supplements the other. Light detection and ranging (LIDAR) is another powerful that can be used in both horizontal and vertical applications to create 3D visualizations.

9.7   LIDAR

Developed in the 1960s, LIDAR itself is not new. More recent development of accurate inertial navigation systems and global positioning systems has allowed the deployment of current LIDAR systems capable of high precision and accuracy to measure water and cloud vapor (Ackerman, 2003). An inertial navigation system (INS) controls aircraft pitch, roll, and yaw, while a GPS is used for flight path and altitude navigation and control. In the case of building a DEM, the GPS provides the means to navigate flights that edge-match without large amounts of overlap, thus increasing quality while maintaining cost-efficiency. Earlier laser-based instruments consisted of single laser pulses.

Current LIDAR technology involves the rapid scanning of 10,000 to 15,000 pulses per second in a pattern perpendicular to an airplane flight path. These pulses are emitted from the aircraft toward Earth's surface and reflect from the surface and objects on the surface back to the aircraft. The rate of pulse emission, speed of the aircraft, and altitude all play an important role in LIDAR mapping. These roles can be thought of similar to those of conventional aerial photography, where scale can be determined based on aircraft altitude and camera focal length. However, in the case of LIDAR, emission angle must be determined since the scanner emits laser pulses and acquires them from multiple angles from a nadir (i.e., a point perpendicular to the emitter toward Earth's surface). LIDAR has been used in the study of atmospherics, floodplains, forest canopy density, agriculture, and geology (Strikon et. al., 1998). By far, most LIDAR applications are for construction of DEM. When building a DEM, a series of raw data points are generated for a landscape. The LIDAR instrument is calibrated prior to beginning operation through alignment with predetermined ground control points. The INS is also calibrated at this time. These calibrations and the initial raw data are compared, and accuracies of 50 cm or less are common.

It should be remembered that all landscape objects are visible at this stage. These objects can be filtered from the initial images, thus allowing for the construction of a DEM. Further interpolation of the DEM that depends on either captured LIDAR data points or ground-collected elevation points may be included in the data set. The filtered data containing objects is kept and may be analyzed: for example, when investigating forest cover types, constructing 3D visualizations or delineating routes, linework, or other physical attributes. The portability and rapid turnaround of LIDAR information has led to the study of ice sheets in Canada, aerosols in the city of Berlin, Germany, and earthquake-prone areas in the Seattle, Washington area. In emergency or natural disasters, where conditions change rapidly, LIDAR can be used effectively for the purposes of monitoring change.

Since the data can be acquired locally and processed locally, quicker management decisions can be made and observation of changing conditions investigated more closely. In the case of natural disasters, LIDAR has other benefits, due to being an aerial application, therefore reducing need and possible injury to ground-based monitoring. Similarly, environments where aerosols or other dangerous airborne pollutants are present can be monitored from a distance. To determine the shapes of objects and their corresponding heights, the points returned to the emitter are collected and filtered. This can be done a number of ways, including:

- Density of points with similar range and return times
- Density of points using neighborhood or proximity analysis
- Correlation to other information, including aerial photographs
- Spectral analysis in the case of multispectral emitters
- Object analysis linking physiology and biology
- Structural analysis of building design and surfaces

In each of these cases we are interested in analysis of the returned light pulses. There is considerable current research into the development of algorithms that can be used to correlate landscape objects and environments to LIDAR-returned point distribution, both spatially and temporally. These investigations often involve methods and techniques similar to those used in remote sensing analysis. LIDAR results in a coarse 3D image with very high precision and accuracy. The surfaces are in 3D and topology is built for them. The images are not photo-realistic in the sense of features having identifiable surfaces. Instead, *textures,* digital images of a realistic surface, are applied to the surfaces from within GIS or a visualization software package. For example, if the visualization is to appear photo-realistic, images are taken of the surfaces of buildings, roads, and paths. These real depictions are then applied (pasted) to the polygons of a LIDAR surface. Alternatively, an aerial photograph or satellite image can be used as a texture and applied to a LIDAR or other continuous surface. In this case the DEM itself and the image being draped or applied as a texture will require that both are georeferenced. Displaying or visualizing a LIDAR surface in 2D defeats the purpose of acquiring the data. Although much more accurate, LIDAR information is also usually more expensive.

But the overriding issue with LIDAR data is that they are in 3D; thus applications and visualization that capitalize on that topology are preferred. LIDAR is closely linked to laser imaging from the side. Portable laser scanners are available which can be set up on a tripod to scan gas plants, building exteriors, construction, and other sites, building a 3D model of them quickly. There are several advantages to using a laser scanner, including their portability, speed, and high accuracy. To acquire the same images using alternative means would require having access to the architectural drawings and then generating the structures in 3D. Textures must also be applied to surfaces derived from laser scanners.

Usually, satellite and aerial photographs are not useful for this purpose, due to the fact that they are captured from above, whereas a laser-scanned image is derived from the side or has an oblique viewpoint. This then necessitates taking several digital images of the structures and pasting them as textures to individual polygons. Currently, a significant number of researchers internationally are studying natural environments and landscapes with a view to developing algorithms derived from LIDAR data that recognize common landscape attributes.

9.8 THREE-DIMENSIONAL ENVIRONMENTS

Three-dimensional environments are those environments that render or project images providing a sense of depth. Objects in a 3D environment can usually be manipulated and respond to the user. The ability to respond to the viewer may be either high or low and is referred to as *interactivity*. Higher levels of interactivity engage the viewer to participate in the visualization more fully. Viewers can interact with the environment in a multitude of ways, including tracking cameras that monitor their movements or respond to their head movements, adjusting the visualization in response. Alternatively, the visualization can zoom in or out, turn or twist, as it responds to a remotely controlled device held in the palm of the viewer. These devices are referred to as *trackers*.

A viewer uses a tracker to interact with a 3D environment. The greater the number of responses possible using the tracker, the higher the level of interaction. Other trackers might include feet movement, sound movement, or pressure sensors responding to a tightening grip as the viewer stands within the environment. Most 3D environments attempt to surround the user, that is, provide a field of view (FOV) that encircles the user. Consequently, the viewer feels immersed in the visualization—thus the term *immersive environments*. These may include rooms with enclosed walls, open walls, or spherical domes. In a 3D environment there may be varying levels and types of representation. Objects will appear in 3D but may not be seen accurately with respect to scale as they are turned or manipulated. Many images and content from alternative media may be viewed in a 3D environment, including wind flows, geology, buildings, parks, hydrology, oil and gas facilities, climate changes, and many more. There is no restriction on content that can be used in a 3D environment in that the image itself is usually 3D, that is, has a 3D topology.

The simplest way to put an image into these types of environments is through the use of VRML. Visualizations seen in such environments have crossed from GIS into the visualization realm and are more closely associated with visualization software. By this we do not mean that content produced from GIS cannot be rendered in 3D visualization environments, only that GIS software has not yet been fully designed for immersive environmental types of applications. GIS software is designed for GIS and, as mentioned earlier, originates from the database. Ideally, the goal would be a visualization environment with GIS functionality, including interactive database access, while maintaining superlative graphics. That is why we state that GIS and visualization originate from two different realms.

Since 3D environments are complex, they require high-end computing hardware, using supercomputers in most cases. Focus is on speed and rendering many millions of polygons per second to maintain high levels of real-time interaction. A 3D visualization must be able to move and change quickly, responding to the viewer at the speed for which the viewer engages the visualization. Often, scales are not apparent in 3D visualizations, and a lot of generalization does take place as the viewer zooms in and out rapidly. This then means that these virtual environments need a mechanism that enables the viewer to understand these elements as the visualization changes perspective and responds. Including cartographic representation and elements in a visualization environment is a challenge. In some cases a single screen or split screens or even numerous smaller screens, with each portraying different information useful to understanding the visualization, have been used.

When two or more screens of information are presented, we have *multirepresentation*. Individual screens or smaller screens may be embedded within larger images, which show, for example, top, side, and perspective views. Or, smaller tables or charts may be present which provide text information associated with the objects being rendered. Each of these is designed to orient the viewer with the image presented. There is a limit to this, it would seem, since the mind and the eyes quickly become saturated with too much information. Therefore, one concern about immersive environments is that they need to display in some way the minimal amount of useful information possible in such a way that the viewer engages the rendering more fully, that is, with greater interactivity.

It is interesting to talk to people who have experienced a 3D environment. Think about the last time you visited a stereo IMAX theatre, where the objects are in 3D. It was probably unlike any other experience you have had viewing 2D images or movies. The objects jump out with depth perception and one feels immediately engaged in the rendering. Updated versions of VRML are being pursued that will directly affect GIS professionals and cartographers interested in using immersive environments. The Web3D Consortium and MPEG-4 are also working toward integration of 3D content with sound and other Web-enabled materials. Geographic markup language (GML) is designed to address issues of coordinates and scales, extending the performance of the VRML standard using extensible markup language (XML). It is GML that has the greatest appeal to geoscience professionals, due to the potential to encode topology, thus opening the door to spatial analysis of Internet cartographic products. GML goals are as follows (OGC, 2001a):

- To provide a means of encoding spatial information for both data transport and data storage, especially in a wide-area Internet context
- To be sufficiently extensible to support a wide variety of spatial tasks, from portrayal to analysis
- To establish the foundation for Internet GIS in an incremental and modular fashion
- To allow for the efficient encoding of geospatial geometry (e.g., data compression)
- To provide easy-to-understand encodings of spatial information and spatial relationships, including those defined by the OGC Simple Features model
- To be able to separate spatial and nonspatial content from data presentation (graphic or otherwise)
- To permit easy integration of spatial and nonspatial data, especially for cases in which the nonspatial data is XML encoded
- To be able to link spatial (geometric) elements readily to other spatial or nonspatial elements
- To provide a set of common geographic modeling objects to enable interoperability of independently developed applications

GML could become widely used because of the benefits and ability to share and gather data from numerous sources. For example, in the United States, the U.S. Department of Agriculture through its Common Computing Environment (CCE) initiative, is placing several thousand computers, acting as servers, throughout the United States, to enable agricultural producers to acquire various sorts of information, including geographic information, useful for agricultural management and operations. This could have a considerable impact on GPS and satellite image providers as well as a host of other geotechnology data providers. Under this scenario, raw or processed GPS data and DGPS corrections could be shared and linked through the Internet.

A producer might also input data to the system and have the data analyzed using the processing power of several parallel computers. Information would be visualized and transferred using XML and GML together. Thus, the concept of visualization will not only be based on going to high-end laboratories or facilities, but will arrive in the user's home or workplace in a format suitable for GIS analysis and viewing. SGML, international standard metalanguage (a language for describing other languages), for text markup systems is designed to allow the exchange of information on any level of complexity among a variety of software and hardware products. XML (W3C) is a shorter version of SGML. It is established as a successor to HTML for better modeling of text and other kinds of structured information, for ease in writing and understanding applications, and for the delivery and interoperability of information between different computing systems and over the Web. Visualization of geographic information is becoming very common using the Internet. Each of these standards is attempting to include the unique functionality of GIS: datums, scalability, interoperability, and improved speed.

9.9   FOUR-DIMENSIONAL VISUALIZATION

Four-dimensional visualization includes (x,y,z) coordinates as well as time. The time component may be real time, where the visualization is updated in a continuous fashion from minute to minute. It is often difficult to determine the true meaning of real time since computing systems must process, store, and then render information to media, although any visualization that updates for more than 1 hour would probably not be called a real-time visualization. The time component is unique to geotechnologies, due to the fact that so many geotechnologies are capable of encoding or time stamping data. In this way data are assigned a time for the moment that the data are captured.

The (x,y,z) coordinates gathered using GPS are an example of this, and the time can be checked for each coordinate using mission-planning software. We might call this *data time* since it relates to the collection of data. Alternatively, there may be a presentation or rendering time, which could indicate change over time. A series of renderings, layers, and themes in GIS that depict change over time would be an example. If enough individual images of themes were tied together, they would form an animation. If the animation had high resolution and changed for short durations, that would be a movie.

Currently, many geotechnologies are being integrated toward providing animations or movies suitable for both immersive environments but also suitable for desktop applications. Usually, these are in the form of MPEG, Quicktime, or other formats. The shift from 2D to 3D to arrive at 4D requires increasingly higher levels of memory and processor speed. It is not unusual for a 4D representation in OpenGL to tax a desktop processor significantly if the rendering is large and the time change duration between events is fine enough. Once the rendering is projected in full frame size and textures and other visualization effects applied, the computer must be capable of processing all these elements rapidly.

Delays in being able to render rapidly reduce user interaction and decrease the sense of realism. It is difficult to determine exact processors and computer configurations that will work as compared to those that will labor in presenting visualizations since 4D visualizations depend on numerous factors:

- Size of the rendering
- Resolution of the rendering
- Computer processor speed

- Hard drive access rate
- Monitor refresh rate
- Sampling time step
- Telecommunications connections (if involved)
- Use of image encoding and compression technologies

But if one is considering a high-end computer for GIS visualization, he or she is advised to make a CDROM of the information and to try out a few computers to learn what results can be expected. For 4D visualizations, 3D objects are often used. They may be created in a 3D modeling package, then imported into the visualization. Textures, effects, and brushes and other visual functionality varies among visualization packages. For those interested in landscapes and environmental content, real images of natural vegetation can be used to increase the sense of realism. This includes textures such as sidewalks, cement, bricks, and building exteriors.

EXERCISES

9.1. Explain what *visualization* means.

9.2. Discuss how cartography is related to visualization, outlining issues with respect to accuracy.

9.3. Discuss how GPS accuracy affects visualizations. Provide examples for 2D and 3D data acquired using points, lines, and polygons.

9.4. Discuss data resolution versus rendering resolution. What are the considerations in rendering real-time geographic information as visualizations?

9.5. Four-dimensional visualizations include time. Explain the significance of time in visualizations and include briefly why you do or do not think including time can result in increased communication.

9.6. Since GIS originates from the database and visualization from the camera, discuss how and where these affect each other.

REFERENCES

Aber, J. 2001. Forest processes and global environmental change: predicting the effects of individual and multiple stressors. *Bioscience,* Vol. 51, No. 9, pp. 738–739.

Aber, J., et al. 2001. Forest processes and global environmental change: predicting the effects of individual and multiple stressors. *Bioscience,* Vol. 51, No. 9, pp. 735–751.

Ackermann, F. 2003. Airborne laser scanning: present status and future expectations. *ISPRS Journal of Photogrammetry and Remote Sensing.* Vol. 54, pp. 64–67.

Anon. 1998. Medical geography: an overview. *Medical Geography Digest,* Vol. 1, No. 2. http://members.nbci.com/mgdigest/digest-2.html.

AUSLIG (Australian Surveying and Land Information Group). 2000. The geocentric datum of Australia. http://www.auslig.gov.au/mapping/docs/250kspec/append_m.pdf.

AUSPOS (The AUSLIG Online GPS Processing System). 2002. Online GPS processing service, Geo-Science Australia. http://www.auslig.gov.au/geodesy/sgc/wwwgps.

Avery, T. E., and Berlin, G. L. 1992. *Fundamentals of Remote Sensing and Airphoto Interpretation,* 5th ed. Prentice-Hall, New Jersey.

Baskent, E. Z. 1999. Controlling spatial structure of forested landscapes: a case study towards landscape management. *Landscape Ecology,* Vol. 14, No. 1, pp. 83–97.

Bergh, J., and Lofstrom, J. 1976. *Interpolation Spaces: An Introduction.* Springer-Verlag, Berlin.

Beutler, G., Gurtner, M., Rothacher, V., and Frei, E. 1989. Relative static positioning with the global positioning system: basic technical considerations. Presented at the IAG/IUGG 125th Anniversary Meeting, Edinburgh, Scotland, Aug. 3–12.

Biezunski, M., Bryan, M., and Newcomb, S., eds. 2000. ISO/IEC 13250 Topic Maps. http://www.isotopicmaps.org/rm4tm/RM4TM-official.html.

Birley, R., and Tasker, N. 1997. Dyscover: a world of special maps for special people. Presented at the 18th ICA/ACI International Cartographic Conference, Stockholm, Sweden, June 23–27.

Blackmore, B. S. 2000. Using information technology to improve crop management. In *Weather and Agro-Environmental Management: Proceedings of the Ag-Met Millennium Conference,* Dublin, Feb. 29, Ireland, pp. 30–38.

Bos, E. S., et al., eds. 1991. *Kartografisch Woordenboek* (Dutch Cartographical Dictionary), Vol. 1.1. Wólters-Noordhoff Atlas producties, Groningen.

British Columbia Ministry of Forests. 2001. Operational field procedures for forest resource survey and mapping using global positioning system technology, V3.0. http://www.for.gov.bc.ca/ric/pubs/teveg/opfield/index.htm.

British National Committee for Geography, Cartography Subcommittee, 1966. *Glossary of*

Technical Terms in Cartography. Royal Society, London. http://www.fes.uwaterloo.ca/crs /geog165/cart.htm.

Bullock, J. B. 1997. GPS placement analysis in an automobile: simulation and performance test results. Presented at the Intelligent Transportation Symposium, Washington, DC.

Burrough, P. 1987. Multiple sources of spatial variation and how to deal with them. In N. Chrisman, ed., *Proceedings of AUTO-CARTA,* Vol. 8, pp. 145–154.

Burrough, P. A., and R. A. McDonnell, 1986. *Principles of Geographic Information Systems for Land Resource Assessment.* Oxford University Press, Oxford.

Burrough, P. A., and Frank, A. U. 1995. Concepts and paradigms in spatial information: are current geographic information systems truly generic? *International Journal of Geographical Information Systems,* Vol. 9, No. 2, pp. 101–116.

Camara, A., and Raper, J. F. 1999. *Spatial Multimedia and Virtual Reality.* Taylor & Francis, London.

Canter, D. 1995. *Criminal Shadows: Inside the Mind of the Serial Killer.* Harper Collins, London.

Clarke, K. C., McLafferty, S. L., and Tempalski, B. J. 1996. On epidemiology and geographic information systems: a review and discussion of future directions. *Emerging Infectious Diseases,* Vol. 2, No. 2, Apr.–June.

Cliff, A., and Haggett, P. 1988. *Atlas of Disease Distributions.* Blackwell, Oxford.

Conroy, M. J., and Noon, B. R. 1996. Mapping species richness for conservation of biological diversity: conceptual and methodological issues. *Ecological Applications,* Vol. 6, No. 3, pp. 763–773.

Cooper, A., and Murray, R. 1992. A structured method of landscape assessment and countryside management. *Applied Geography,* Vol. 12, pp. 319–338.

Couclelis, H. 1992. Geographic knowledge production through GIS: towards a model for quality monitoring. In *Two Perspectives on Data Quality.* National Center for Geographic Information and Analysis, 2001. Santa Barbara, CA.

Craymer, M. R., Milbert, D., and Knudsen, P. 2001. Report of the sub-commission for North America, IAG Commission X (Global and Regional Geodetic Networks), Report to the IAG Scientific Assembly, Budapest, Hungary, June 2–7.

Dodge, M. 2002. The Geography of Cyberspace Directory. http://www.cybergeography.org /geography_of_cyberspace.html.

Douglas, D. H., and Peucker, T. K. 1973. Algorithms for the reduction of the number of points required to represent a digitized line or its caricature. *Canadian Cartographer,* Vol. 10, No. 2, Dec., pp. 112–122.

EC (European Commission). 2001a. Telegeoprocessing. Upcoming conference, TeleGeo'2002. GI and GIS. European Commission, Sophia Antipolis, France. http://lisi. insa-lyon.fr/ ~laurini/telegeo.

EC. 2002. Information Society Technologies: 2002 Work Program. A Program of Research, Technology and Demonstration: 5th Framework Programme. European Commission, Brussels.

EGNOS (European Geostationary Navigation Overlay Service). 1994. France Telecom and DBP Telekom: European proposal for the lease of INMARSAT-III Navigation Transponders, Oct.

EUROGI (European Umbrella Organization for Geographic Information). 2000. *Towards a Strategy for Geographic Information in Europe.* Consultation Paper, Version 1.0.

Executive Office of the President. 1994. Coordinating geographic data acquisition and access the National Spatial Data Infrastructure. Executive order 12906. *Federal Register,* Vol. 59, pp. 17671–17674.

Forman, R. T. T. 1990. Ecologically sustainable landscapes: the role of spatial configuration. In I. S. Zonneveld and R. T. T. Forman, eds., *Changing Landscapes: An Ecological Perspective.* Springer-Verlag, New York, pp. 261–278.

Fortin, M. J., et al. 2000. Issues related to the detection of boundaries. *Landscape Ecology,* Vol. 15, pp. 453–466.

Fu, P., and Rich, P. M. 1999. Design and implementation of the Solar Analyst: an ArcView extension for modeling solar radiation at landscape scales. In *Proceedings of the 19th Annual ESRI User Conference,* San Diego, CA. http://www.esri.com/library/userconf/proc99 /proceed/papers/pap867/p867.htm.

Gahegan, M. 1998. Four barriers to the development of effective exploratory visualization tools for the geosciences. http://www.cs.curtin.edu.au/~mark/visworkshop/visproblems.html.

GeoConnections Policy Advisory Node. 2001. Geospatial data policy study. Project report 03-34257, KPMG Consulting, Ottawa, Canada.

Germs, R., et al. 1999. Multi-view VR interface for 3D GIS. *Computers and Graphics,* Vol. 23, No. 4, pp. 497–506.

Gesler, W. 1986. The uses of spatial analysis in medical geography: a review. *Social Science and Medicine,* Vol. 23; pp. 963–973.

GLONASS (Global Navigation Satellite System). 1999. Decree of the President of Russian Federation of States, Feb. 18. http://www.rssi.ru/SFCSIC/38rp-e.html.

Goodchild, M. F. 1985. Geographic information systems in undergraduate geography: a contemporary dilemma. *Operational* Geographer, Vol. 8, pp. 34–38.

Goodchild, M. F. 1988. Towards an enumeration and classification of GIS functions. In R. T. Aangeenbrug and Y. M. Schiffman, eds., *International Geographic Information System Systems (IGIS) Symposium: The Research Agenda* (3 vols.). National Aeronautics and Space Administration, Washington, DC, Vol. 1, pp. 45–54.

Goodchild, M. F. 1997. What is geographic information science? NCGIA Core Curriculum in GIScience. http://www.ncgia.ucsb.edu/giscc/units/u002/u002.html, posted Oct. 7.

Govindasamy, B., and Caldeira, K. 2001. Geo-engineering Earth's radiation balance to mitigate CO_2-induced climate change. *Geophysical Research Letters,* Vol. 27, No. 14, p. 2141. http://www-pcmdi.llnl.gov/cccm/govindas02.pdf.

Growe, S. 1999. Knowledge based interpretation of multisensor and multitemporal remote sensing images. *IAPRS,* Vol. 32, Part 7-4-3 W6, Valladolid, Spain, pp. 130–138.

GSDI (Global Spatial Data Infrastructure). n.d. http://www.gsdi.org.

Guerrero, I. 2002. *OpenGIS XML—Based Web Map Interfaces.* White paper. Intergraph Corporation, Huntsville, AL.

Gurtner, W. 1996. Guidelines for a permanent EUREF GPS network: In E. Gubler and H. Hornik, eds., *Proceedings of the EUREF Symposium,* Helsinki, Finland, May 1995. EUREF Publication 4, pp. 68–72.

Houyoux, M. R. 1998. Using continuous emission monitoring data for air quality modeling inventories, presented at *Emission Inventory: Living in a Global Environment,* Dec. 8–10, New Orleans, LA, Air and Waste Management Association, Pittsburgh, PA.

Hunter, G. J. 1999. New tools for handling spatial data quality: moving from academic concepts to practical reality, *URISA Journal,* Vol. 11, No. 2, pp. 25–34.

ICA (International Cartographic Association). 1973. *Multilingual Dictionary of Technical Terms in Cartography.* Steiner, Wiesbaden, Germany.

Itami, R. M. 1994. Simulating spatial dynamics: cellular automata theory. *Landscape and Urban Planning,* Vol. 30, pp. 27–47.

Joint Program Office, U.S. Air Force. n.d. Global positioning system. SMC/CZ. http://gps.losangeles.af.mil.

Karl, J. W., et al. 2000. Sensitivity of species habitat-relationship model performance to factors of scale. *Ecological Applications,* Vol. 6, pp. 1690–1705.

Karrow, R. W. 1993. *Mapmakers of the Sixteenth Century and Their Maps: Bio-Bibliographies of the Cartographers of Abraham Ortelius, 1570.* Speculum Orbis Press, Chicago.

Kingsley, G. T. 1999. *Building and Operating Neighborhood Indicator Systems: A Guidebook.* National Neighborhood Partners Institute, Urban Institute Press, Washington, DC.

Kraak, M. J., and Brown, A. 2001. *Web Cartography: Developments and Prospects.* Taylor & Francis, London, p. 87.

Kuiper, J., Ayers, A., Johnson, R., and Tolbert-Smith, M. 1996. Efficient data exchange: integrating a vector GIS with an object-oriented, 3-D visualization system. Presented at the 3rd International Conference/Workshop on Integrating GIS and Environmental Modeling, Santa Fe, NM, Jan.

Lane, N., Assistant to the President for Science and Technology. 1999. Keynote address at the Institute of Navigation GPS'99 Conference, Nashville, TN, Sept. 14. http://clinton3.nara.gov/WH/EOP/OSTP/html/9910_6.html.

Langley, R. B. 1998. The UTM grid system. *GPS World,* Feb., pp. 46–48.

Lodha, S. K., and Verma, A. 1999. Animations of crime maps using virtual reality modeling language. *Western Criminology Review,* Vol. 1, No. 2.

MacEachren, A. M. 1994. Visualization in modern cartography: setting the agenda. In D. R. F. Taylor and A. M. MacEachren, eds., *Visualization in Modern Cartography.* Pergamon Press, London.

MacEachren, A. M., and Kraak, M. J. 1997. Exploratory cartographic visualization: advancing the agenda. *Computer and Geosciences,* Vol. 23, pp. 335–344.

Mamalian, C. A., et al. 1999. *The Use of Computerized Crime Mapping by Law Enforcement: Survey Results.* U.S. Department of Justice. Washington, DC, Jan.

Mariano, G. 2001. Web popularity swells overseas. *CNET News,* May 14. http://news.cnet.com/news/0-1005-200-5921106.html.

Mark, D. 1997. The history of geographic information systems: invention and re-invention of triangulated irregular networks (TINs). *In Proceedings of GIS/LIS'97.* http://www.geog.buffalo.edu/ncgia/gishist/GISLIS97.html.

McHarg, Ian. 1971. *Design with Nature.* Doubleday/Natural History Press, Garden City, NY.

McKercher, R. B., and Wolfe, B. 1986. *Understanding Western Canada's Dominion Land Survey System.* University of Saskatchewan, Division of Extension and Community Relations, Saskatoon, Saskatchewan, Canada.

Molenaar, M. 1998. *An Introduction to the Theory of Spatial Object Modeling for GIS.* Taylor & Francis, London.

Monmonier, M. 1996. *How to Lie with Maps,* 2d ed. University of Chicago Press, Chicago.

Mora, E., Carrascosa, C., and Ortega, G. 1998. *Characterization of the Multi-path Effects on the GPS Pseudo-range and Carrier Phase Measurements.* Institute of Navigation, Nashville, TN.

Murphy, R. R. 1996. Biological and cognitive foundations of intelligent sensor fusion. *IEEE Transactions on Systems, Man, and Cybernetics,* Vol. 26, No. 1, pp. 42–51.

Natural Resources Canada. 2002a. Canada land inventory: GeoGratis. http://geogratis.cgdi.gc.ca/frames.html.

Natural Resources Canada. 2002b. National topographic maps. http://maps.nrcan.gc.ca/topographic.html.

NAVCEN (United States Coast Guard's Navigation Center of Excellence). 2001. *The Global Positioning System Standard Positioning Service Performance Standard.* Assistant Secretary of Defense, for Command, Control, Communications and Intelligence, Oct. http://www.navcen.uscg.gov/gps/geninfo/2001SPSPerformanceStandardFINAL.pdf.

NCGIA (National Center for Geographic Information and Analysis). n.d. Initiative 19: GIS and society: the social implications of how people, space, and environment are represented in GIS. http://www.geo.wvu.edu/il9.

Nyerges, T. L. 1992. Coupling GIS and spatial analytical models. In *Proceedings of the 5th International Symposium on Spatial Data Handling,* Charleston, SC, IGU Commission on GIS, pp. 534–543.

OGC (OpenGIS Consortium). 2001a. *Geographic Markup Language (GML),* 2.0, *Open GIS Specification.* OGC document 01-029. Feb. 20.

OGC. 2001b. Web file serving specification. http://www.opengis.org/info/press/20011102_WFS_RFC_PR.htm.

OGC. 2001c. Geographic markup language (2.0), http://opengis.net/gml/01-029/GML2.html.

Ohtsuka, Y., Fujisaku, J., Nakajima, S., and Hayashi, K. 1997. Development of producing system of tactile map using digital geographic data. Presented at the 18th ICA/ACI International Cartographic Conference, Stockholm, Sweden, June 23–27.

Pan American Health Organization, 1996. Use of geographic information systems in epidemiology (GPS-Epi). *Epidemiological Bulletin,* Vol. 17, No. 1, Mar.

Petrie, G. 2001. The future direction of the SPOT programme. *Geoinformatics, Oct.*

Peucker, T. K., Fowler, R. J., Little, J. J., and Mark, D. M. 1978. The triangulated irregular network. In *Proceedings of the American Society of Photogrammetry: Digital Terrain Models (DTM) Symposium,* St. Louis, MO, May 9–11, pp. 516–540.

Pirsig, Robert. 1974. *Zen and the Art of Motorcycle Maintenance.* Bantam, New York.

Poiker, T. 1999. Spatial modeling. *Course 9 Notes: UNIGIS.* Simon Fraser University, Vancouver, British Columbia, Canada, pp. 29–30.

Poiker, T. K., Moore, P. J., and Berdusco, B. 2000. The education cartogram: educational supply versus occupational demand. *GeoInformatics,* Vol. 3, Sept.

Raper, J. 2000. *Multidimensional Geographic Information Science.* Taylor & Francis, London, pp. 132–133.

Robinson, A. H., et al. 1995. *Elements of Cartography,* 6th ed. Wiley, New York, pp. 61–90.

Rogowitz, B., and Treinish, L. A. 1997. How not to lie with visualization. In *Proceedings of the IBM Visualization Data Explorer Symposium,* San Francisco, Oct. 28–29, 1996.

Scheller, A. 2000. Measuring sustainability: the making of sustainability indicators in interdisciplinary research settings. Presented at the POSTI International Conference: Policy Agendas for Sustainable Technological Innovation, London, Dec.

Schurmann, N. 2001. Critical GIS: theorizing an emerging science. *Cartographica,* Vol. 36, No. 4, pp. 1–108.

Sester, M. 1999. Acquiring transition rules between multiple representations in a GIS: an experiment with area aggregation. *Computers, Environment and Urban Systems,* Vol. 23, No. 1, pp. 5–17.

Shirley, R. W. 1983. *The Mapping of the World: Early Printed World Maps, 1472–1700.* Holland Press, London.

Sinton, D. 1978. The inherent structure of information as a constraint to analysis: mapped thematic data as a case study. In G. Dutton, ed., *Harvard Papers on Geographic Information Systems,* Vol. 7. Addison-Wesley, Reading, MA.

Slaymaker, D. M., et al. 1996. Mapping deciduous forests in Southern New England using aerial videography and hyperclustered multi-temporal Landset™ imagery. In J. M. Scott, T. H. Tear, and F. Davis, eds., *Gap Analysis: A Landscape Approach to Biodiversity Planning,* American Society for Photogrammetry and Remote Sensing. Bethesda, MD.

Slocum, T. A. 1999. *Thematic Cartography and Visualization.* Prentice Hall, Upper Saddle River, NJ.

Snyder, J. P. 1982. *Map Projections Used by the U.S. Geological Survey,* 2nd ed. U.S. Government Printing Office, Washington, DC.

Snyder, J. P. 1993. *Flattening of the Earth: Two Thousand Years of Map Projections.* University of Chicago Press, Chicago.

Spanner, M. A., L. L. Pierce, S. W. Running, and D. L. Peterson. 1990. The seasonality of AVHRR data of temperate coniferous forests: relationship with leaf area index. *Remote Sensing of Environment,* Vol. 33, p. 112.

Star, J., and Estes, J. 1990. *Geographic Information Systems: An Introduction.* Prentice Hall, Upper Saddle River, NJ, pp. 2–3.

Stein, M. L. 1999. *Interpolation of Spatial Data: Some Theory for Kriging.* Springer-Verlag, New York.

Strickon, J., Rice, P., and Paradiso, J. 1998. Stretchable music with laser rangefinder. In *SIGGRAPH '98 Conference Abstracts and Applications.* ACM Press, New York, p. 123.

Tansley, A. G. 1914. Presidential address. *Journal of Ecology,* Vol. 2, pp. 194–203.

Thompson, S. K. 1992. *Sampling.* Wiley, New York.

Thrower, N. J. 1991. When mapping became a science. *UNESCO Courier,* June, p. 31(4).

Thurston, J. 2001b. Vertical GIS and visualization: a comparison of 2-D and 3-D. M.Sc. thesis, Manchester Metropolitan University, Manchester, Lancashire, England.

Thurston, J., Moore, P., Noftsker, C., Poiker, T., and Parkinson, B. (2001a). Visualization and GIS. *GeoInformatics,* Vol. 4, No. 1, p. 41.

Thurston, J., and Pruss, S. 2000. Geo-temporal and geo-spatial relationships of coyote populations in Elk Island National Park, Canada. In *Proceedings of GIS 2000,* Toronto, Ontario, Canada, Mar. 13–16.

Thurston, J., Moore, J. P., Parkinson, B., and Poiker, T. K. 2001b. Visualization and GIS. In *Proceedings of GIS 2001,* Vancouver, British Columbia, Canada.

Tobler, W. 1970. A computer movie simulating urban growth in the Detroit region. *Economic Geography,* vol. 46, No. 2, pp. 234–240.

Tomlin, D. 1990. *Geographic Information Systems and Cartographic Modeling.* Prentice Hall, Upper Saddle River, NJ, p. 11. Reprinted in *American Geographers,* Vol. 87, No. 2, pp. 346–362, 1997. http://dusk.geo.orst.edu/annals.html.

Tukey, J. W. 1977. *Exploratory Data Analysis.* Addison-Wesley, Reading, MA.

Turvey, B. E. 1999. *Criminal Profiling: An Introduction to Behavioural Evidence Analysis.* Academic Press, London.

UCGIS. (University Consortium for Graphic Information Science). 1997. GIS history project. Presented at the UCGIS Summer Assembly, Bar Harbor, ME, June. http://ucgis.org.

UNAVCO (University Navstar Consortium). n.d. Metadata and cut/splice (time) editing. http://www.unavco.ucar.edu/data_support/software/edit/edit.html.

U.S. Department of Commerce. 2000. Civilian benefits of discontinuing selective availability. Fact sheet. May. http://osecnt13.osec.doc.gov/public.nsf/docs/852E6E9C5E9E90F78525-68D3005BEFC8.

USGS. 2002. The national map-orthoimagery. http://mac.usgs.gov/mac/isb/pubs/factsheets/fs10702.html.

Unwin, K. I. 1975. The relationship of observer and landscape in landscape evaluation. *Transactions of the Institute of British Geographers,* No. 66, pp. 130–133.

van der Schee, L. H., and Jense, G. J. 1995. Interacting with geographical information in a virtual environment. In *Proceedings of JEC-GIS,* The Hague, The Netherlands, Vol. 1, pp. 151–157.

Walker, R., ed. 1993. *AGI Standards Committee GIS Dictionary.* Association for Geographic Information, London.

Wilford, J. N. 2002. *The Mapmakers: The Story of Great Pioneers in Cartography—From Antiquity to the Space Age.* Pimlico, Random House, London.

Wilson, A., and Wilson, J. 1976. *The Making of the Nuremberg Chronicle.* Amsterdam.

Wood, M. 1972. Human factors in cartographic communication. *Cartographic Journal,* Vol. 9, No. 2.

Wormley, S. J. 2003. Sam Wormley's global positioning systems (GPS) resources. Educational Observatory Institute. http://www.edu-observatory.org/gps/gps.html.

Wright, D. J., Goodchild, M. F., and Proctor, J. D. 1997. Demystifying the persistent ambiguity of GIS as "tool" versus "science." *Annals of the Association of American Geographers,* Vol. 87, No. 2, pp. 346–362.

INDEX